WOMEN

—— AND THE ——

ART OF WAR

SUN TZU'S STRATEGIES
FOR WINNING WITHOUT
CONFRONTATION

Catherine Huang ★ A.D. Rosenberg

TUTTLE Publishing

Tokyo | Rutland, Vermont | Singapore

The Tuttle Story: "Books to Span the East and West"

Most people are surprised to learn that the world's largest publisher of books on Asia had its beginnings in the tiny American state of Vermont. The company's founder, Charles E. Tuttle, belonged to a New England family steeped in publishing. And his first love was naturally books—especially old and rare editions.

Immediately after WW II, serving in Tokyo under General Douglas MacArthur, Tuttle was tasked with reviving the Japanese publishing industry, and founded the Charles E. Tuttle Publishing Company, still thrives today as one of the world's leading independent publishers.

Though a westerner, Charles was hugely instrumental in bringing knowledge of Japan and Asia to a world hungry for information about the East. By the time of his death in 1993, Tuttle had published over 6,000 books on Asian culture, history and art—a legacy honored by the Japanese emperor with the "Order of the Sacred Treasure," the highest tribute Japan can bestow upon a non-Japanese.

With a backlist of 1,500 titles, Tuttle Publishing is more active today as at any time in its past—inspired by Charles' core mission to publish fine books to span the East and West and provide a greater understanding of each.

Published by Tuttle Publishing, an imprint of Periplus Editions (HK) Ltd.

www.tuttlepublishing.com

Library of Congress Cataloging-in-Publication Data

Huang, Catherine.
  Women and the art of war : Sun Tzu's strategies for winning without confrontation / by Catherine Huang and A.D. Rosenberg. -- 1st ed.
     p. cm.
  ISBN 978-0-8048-4254-9 (hbk.)
1. Businesswomen--Psychology. 2. Success in business. 3. Career development. 4. Military art and science. 5. Sunzi, 6th cent. B.C. Sunzi bing fa. I. Rosenberg, Arthur D. II. Title.
  HD6054.3.H83 2011
  650.1082--dc23

                              2011019341

ISBN 978-0-8048-4169-6

Distributed by

North America, Latin America & Europe
Tuttle Publishing
364 Innovation Drive
North Clarendon, VT 05759-9436 U.S.A.
Tel: 1 (802) 773-8930; Fax: 1 (802) 773-6993
info@tuttlepublishing.com
www.tuttlepublishing.com

Japan
Tuttle Publishing
Yaekari Building, 3rd Floor
5-4-12 Osaki, Shinagawa-ku, Tokyo 141 0032
Tel: (81) 3 5437-0171; Fax: (81) 3 5437-0755
sales@tuttle.co.jp
www.tuttle.co.jp

Asia Pacific
Berkeley Books Pte. Ltd.
61 Tai Seng Avenue #02-12
Singapore 534167
Tel: (65) 6280-1330; Fax: (65) 6280-6290
inquiries@periplus.com.sg
www.periplus.com

15 14 13 12 11   6 5 4 3 2 1   1111RP

Printed in China

TUTTLE PUBLISHING® is a registered trademark of Tuttle Publishing, a division of Periplus Editions (HK) Ltd.

Contents

INTRODUCTION

"Conquer yourself and the world will lie at your feet."

The gender gap is closing in North America and some other countries around the globe. But even here the gap is still wide enough to frustrate and stifle far too many women seeking an equal opportunity to improve their lives and realize their ambitions.

Women rightfully seek an even playing field. Many men with genuine self-confidence and a sense of fairness may share this imperative, but reality suggests that the majority do not. The reason, we believe, is that men and women view themselves, their environments and the world differently. We will address some of these differences as they apply within the context of this book.

Male dominance has always been the way of the past, but history is a poor justification for clinging to inequities rooted in cave-man images and old-fashioned discrimination. While many women have the support of husbands, fathers, brothers and friends, change is slow — and no blame to women for being tired of waiting for our society to wake/ grow up.

Fortunately, there is no need to set the world on fire, only to light a flame beneath a few stubborn posteriors that may be cluttering your path.

The Book

"All warfare is based upon deception." –Sun Tzu

Sun Tzu's *The Art of War* dates back to the 6th century BCE. It is widely considered to be the earliest and most universal treatise ever written on the basic principles of warfare. Its central theme: the conducting and winning of war with *minimum* confrontation, risk and loss. This book, along with other military classics, has been widely studied and embraced as a primer for devising business strategy throughout Asia and the US.

Su-Tzu's contention that *"All warfare is based upon deception"*(allowing your enemy to perceive you as weak) seems to imply that peace has to be founded on honesty and openness—in other words, the absence of "art." Interestingly, the word westerners have traditionally translated as "art" more precisely translates into "as water goes." This idea of a natural path relates to the precepts and methodologies found in the *Tao Te Ching** , the influence of which will be referenced several times within the pages of this book.

Given the period and purpose of this work, the Sun Tzu's arrangement of chapters and topics doesn't always offer a one-to-one correlation with more peaceful pursuits. However, his strategies and tactics for achieving success with maximum efficiency and minimum expense remain applicable to any number of modern activities that involve competing and coming out on top.

This is not to say that Sun Tzu's ideas are terribly complex, only that they aren't necessarily organized and presented in a way that's easy to apply. For example, many of his suggestions are intended to be used in concert rather than as individual actions. It follows that a single change can alter an entire strategy.

---

*   Also *Dao De Jing*: a 6th century BCE classic on Taoism attributed to Lao-tzu.

Consistent with the teachings of the Tao, which clearly influenced the writer, there are no hard-set rules to memorize. In this school of thought, extremes like good and evil are closely related, as each exists only in terms of the other. Similarly, the way a plan worked for you yesterday can fall short tomorrow. This is why we focus on an understanding of the principles taught in *The Art of War*, rather than a list of rights and wrongs. For example, references to the "enemy" may correspond to your competition for a promotion or benefit. In situations where men enjoy a gender-based advantage, the enemy may be the challenges a woman needs to overcome in order to attain a favorable playing field.

The number of self-improvement books sold each year seems to suggest that millions of people are taking greater strides to increase their chances of changing themselves and their circumstances for the better. Improvement, however, is a multi-step process: first comes acquiring knowledge and understanding what you've read, watched or listened to. Then, you must apply it to real-life situations. This is to say that learning how to do something isn't enough — you also have to *use* it in a practical and meaningful way.

Applying your new knowledge helps to build confidence, which can lead to more success and increased assurance. It's no secret that winning breeds success in any environment.

How The Art of War Relates to Women
Being successful as a woman doesn't have to mean beating men at their game. Many women have discovered the advantages of competing on their own terms. We've all heard the clichés about women needing to work twice as hard and achieve three times as much as men in order to receive half the recognition. Well and good, but it's time to put these catchy maxims aside and focus on what works.

Applying Sun Tzu's *The Art of War* to women's issues, or to any other pursuit, is something of a metaphor. Fortunately, exploiting women's strategies and tactics need not require littering the path with wounded egos. In fact, the time-honored female wisdom of avoiding direct confrontation is very much in keeping with the teachings of Sun Tzu. Since the dawn of civilization, women have recognized the folly of wasting energy and resources on unproductive posturing and conflict. That's why Sun Tzu's efficient and practical advice is so well suited to women's objectives.

Another point to consider is the larger picture of what winning really means. Let's face it: keeping women down and out of leadership and decision-making is a loss to everyone! The predominantly male practice of winning at the loss of someone else (*win-lose*) may provide some temporary gratification, but a *win-win* structure serves both parties' interests in the longer run. And since most women are aware of this, it remains a goal worth working toward.

That said, there are times when confrontation cannot entirely be avoided in the pursuit of one's legitimate interests. To this end, Sun Tzu shows us how to compete by emphasizing our unique individual and collective strengths.

Who was Sun Tzu?
First of all, "*tzu*" in this context essentially means "Master," a term used to honor a renowned teacher or philosopher. According to Chinese scholars, *Sun Wu* (his actual name) was an advisor to the King of Wu (no relation) toward the end of China's Spring and Autumn Period (722–481 BCE).

Born around 515 BCE, Sun Wu was the son of a warrior and grandson of a general. His family background gave him privileged access to various books and documents on warfare. As a youth he came to the kingdom of Wu to escape turmoil in his native Qi. Over the years—some twenty or more — he became a keen observer of life and combat. His written observations became the treatise on waging war called *Bing Fa* (*Soldier's Skill*), known to us as *The Art of War*.

Story has it that during this time, the King of Wu was looking for a commander-in-chief to head an attack on a neighboring state. Sun Wu applied for the job, sharing with the king his writings on organization and strategy in an attempt convince him that he was qualified. He agreed to submit to a test of his tactics by organizing more than 300 royal concubines and maids into a well-drilled unit. The women maintained that they understood Sun Wu's directions, but laughed when he gave his commands. Obedience to a clearly given order being the responsibility of the officers, Sun Wu ordered the beheading of two of the troop leaders and appointed two others in their place. The new unit took orders more seriously thereafter.

Sun Tzu's military successes are legendary. In a famous battle against the Kingdom of Zhou, he defeated an army of 200,000 soldiers with a force one-tenth the size. According to the 2nd century BCE biographer, Sima Qian, Sun Tzu enjoyed a successful military career and may have later updated *The Art of War* based upon his personal experience.

Sun Tzu was about winning efficiently, with a minimum of cost and waste. If this brilliant tactician were alive today and living in the West, what advice might he offer to women in their attempts to level the playing field for opportunity and recognition? Allow us to speculate that he'd have advised women of all ages to familiarize themselves with *The Art of War* and adapt it to their particular circumstances and needs.

This is what your authors humbly offer in *Women and The Art of War*.

The Art of War

"Warfare is the greatest affair of state.." – *Sun Tzu*

The following version of *The Art of War* is based upon the public domain English language translation by Lionel Giles, but one that has been respectfully and thoroughly edited by your authors. Our intent is not to create a more literal translation but, rather, one that may be more accessible to our western audience.

PART ONE

The Art of War

1. Planning

Sun Tzu wrote: The art of war is vitally important to the interests and existence of the State. In fact, it is the very basis of life and death and the way to survival or extinction. For these reasons, is necessary to comprehensively analyze and consider it.

Key Factors

To this end, the art of war should be evaluated and structured in terms of the following five factors:

1. Legal and moral standards (according to the Tao)
   The ruler's subjects should follow his authority, by which they willingly live and die for him without fear.
2. Nature (weather and related conditions)
   Heaven and Earth contain yin and yang such as night and day, hot and cold, and seasonal variations.
3. Earth (terrain)
   Terrain refers to near and far, hard and easy, and various shades of safe and unsafe.
4. Leadership
   A good general (leader) possesses the qualities of wisdom, credibility, consideration, courage and discipline.

5. Organization and discipline
   Essential elements in leading people and managing logistics.

Every general (leader) should be familiar with these basic elements. Those who are will be successful; those who aren't are destined to fail.

Comparisons

When seeking to determine military conditions, use the following issues as your basis of comparison:

1. Which of the two rulers (ours or theirs) best complies with legal and moral standards?
2. Which leader appears to be more capable?
3. Which group holds the greatest advantages in natural conditions and terrain?
4. Which side is better disciplined?
5. Whose resources are stronger?
6. Which side is better trained?
7. Which side is more consistent and fair in giving out rewards and punishment?

This evaluation will enable me to predict victory or defeat. The general who follows and acts upon this advice will win and should be kept in charge. The one who neither pays attention to my counsel nor acts upon it will lose, for which reason you should replace him.

Taking all of this into consideration, seek any additional advantages that may be available. Mold your tactics to the existing external factors and follow a flexible strategy that compensates for any tactical imbalances.

Deception

Sun Tzu tells us here (and continuously throughout *The Art of War*) that warfare is based upon deception. Thus he advises us to appear (to the enemy) weak when we are strongest and:

1. Pretend to be resting while you advance;
2. When you are far away, create the appearance of being near, and when you're close, feign being far away;
3. Use bait (e.g., the illusion of a weakness) to tempt the opposition;
4. Confuse the enemy with false signs and information;
5. Prepare for a substantial enemy but avoid them if they are very strong;
6. If they appear flustered or angry, try to irritate them;
7. If they're resting, force them to use their energy;
8. If they are cohesive, introduce conflict into their midst;
9. Attack where they are least prepared;
10. Move forward when they don't expect it.

These are the strategies that lead to success. They should not be divulged in advance even to your own side, and must be designed as opportunities present themselves.

The general who wins a battle will have meticulously calculated his plans beforehand. Thorough preparations promote victory, whereas indifferent calculations increase the likelihood of defeat.

2. Waging War

Sun Tzu wrote: In order to field a thousand swift chariots, an equal number of heavy chariots, and a hundred thousand mail-clad soldiers, with enough armor and other items (such as glue and paint) to carry them a thousand li*, the total expenditure will amount to a thousand ounces of silver per day.

---

* Approximately 500 kilometers, or 310 miles today (a bitt less in ancient times)

Seek a Timely Victory

"I have heard of awkward haste, but have never observed skill in overlong campaigns." – Sun Tzu

When you engage such a large force in battle, a prolonged campaign will dull your men's weapons, dampen their enthusiasm and deplete your funds. If you attack cities, the men's strength will be exhausted. Then the enemy will rise to take advantage of your weakened forces, and even the wisest leader will be unable to control the consequences. Only those who are well acquainted with the dangers inherent in employing a major force are capable of truly understanding how to engage a military action to advantage.

Provisions

A skillful leader does not conscript the same people more than once or transport provisions more than twice. At first, bring your equipment from home, and then forage upon the enemy. This should be enough to feed your army. Continuing to transport food and other provisions from home will impoverish the state.

Those who are in close proximity to the army will raise their prices, causing the people's substance to be drained away. This further leads to strains upon the rich (who are supporting the army) and the poor, who will not be able to afford adequate food and living necessities. The expenses of the rich will amount to seven tenths of their wealth, and the ruler's unrecoverable expenses for ruined chariots, broken-down horses, damaged breast-plates, helmets, bows and arrows, spears and protective shields, sturdy oxen and heavy wagons, will expend six-tenths of his total resources.

Thus the wise general will forage on the enemy. A single bushel of the enemy's food is worth twenty of his own, and the same applies to a picul* of fodder.

---

* Slightly over 133 pounds (about 60 kilograms).

Motivation

Anger is what motivates soldiers to slay the enemy, and material rewards encourage them to seize the enemy's property. In a fight where ten or more chariots have been captured, give a reward to the first man who captured one. Then remove the enemy's flags and insignias from the captured chariots, replace them with your own, and use them alongside ours.

Treat the captured soldiers well, and reorient those you can to fight on your side. This is known as conquering the enemy to strengthen your own side.

To summarize, focus your objective on victory, not a prolonged siege. The general who understands warfare is thus the master of the people's fate, and is responsible for the safety or endangerment of the nation.

3. Attack Strategically

Sun Tzu wrote: Overall, the best way to defeat your enemy is to take their state undamaged. This is highly preferable to destroying it. It follows that capturing their forces is better than annihilating them, all the way from the entire army down to its battalions, companies and squads.

Thus winning a hundred victories is not the height of excellence; breaking their resistance without fighting is the true height of excellence.

Strategy

The most efficient and effective warfare policy is to neutralize the enemy's plans; next to block their alliances; then to attack their army; and last (worst) to siege their fortified cities.

As a rule, attacking fortified cities should be avoided whenever possible and undertaken only as a last resort. Building massive protective shields capable of being moved forward, not to mention armored assault vehicles and a host of additional materials, will take three whole months;

and building mounds of earth against the walls will take another three months to finish.

Meanwhile, the general may not be able to control his impatience and may launch his men into the assault like swarming ants. In such a case he will lose one-third of his force, and the city will remain untaken. Such are the disastrous effects of a siege.

The skillful leader subdues the enemy's troops without fighting; he captures their cities without laying siege to them, and overthrows their realms without a prolonged battle. His objective is to exercise his mastery while preserving his forces intact, leaving his weapons sharp and enabling him to preserve his military gains. Such is the strategy of a successful campaign.

Following the Odds
In most cases, the best strategy for deploying troops is the following:
1. If your strength equates to ten to the enemy's one, surround them;
2. If five to one, attack them;
3. If two to one, divide your army into two separate forces;
4. If your strength is equal to theirs, attack when other circumstances favor you;
5. If you are slightly inferior in numbers, protect yourself and seek to circumvent the enemy;
6. If you are overmatched, avoid them.

Although a small force may fight well, a lack of flexibility will result in its being captured by the larger force.

The Wrong Stuff
The general is the pillar of the state. If he is strong and competent in all regards, the state will be likewise strong; if the pillar is defective, the state will weaken. Either way, the

ruler of the state can create difficulties for the military in three specific manners:

1. By commanding the army to advance or retreat without understanding their position and circumstances. This is called hobbling the army.
2. By attempting to govern an army in the same way as he administers a kingdom, with no knowledge of the conditions that affect the army. This confuses and unsettles the officers.
3. By issuing commands to the army despite ignorance of military principles and tactical circumstances. This makes the officers doubtful.

Confusion and doubt lead destined to invite trouble from neighboring rulers. Such anarchy will open the door to an opponent's victory.

The Right Stuff
These five essential factors will assure victory to the side whose general:

1. knows when to fight and when not to fight;
2. knows how to deploy both superior and inferior forces;
3. whose army is animated by the same spirit throughout all its ranks;
4. has prepared himself and waits to take the enemy unprepared;
5. is capable and is not interfered with by the ruler.

Thus it is said that if you know:

1. the enemy and yourself, you need not fear the result of a hundred battles;
2. yourself but not the enemy, for every victory you will also suffer a defeat;
3. neither the enemy nor yourself, you will be defeated in every battle.

4. Military Disposition

Sun Tzu wrote: In ancient times, the skilled warriors protected themselves — indeed, made themselves invulnerable --while awaiting an opportunity to defeat the enemy.

Securing yourself against defeat lies in your own hands; the opportunity to defeat the enemy is provided by the enemy himself. Thus the able fighter can secure himself against defeat, but cannot be certain of defeating the enemy.

Hence the saying: You may know how to conquer the enemy, but the time may not be ripe to implement your plan.

Defense/Offense

If conditions do not favor an attack, you must set up your defense; when you can defeat the enemy, then go on the offensive. A defensive posture implies insufficient strength; attacking suggests abundant strength.

The general who is skilled in defense hides in the most secret recesses of the earth; he who is skilled in attack strikes forth from the upper heights of heaven. Thus on the one hand they protect themselves, and on the other achieve a complete victory.

Excellence

To view a victory only when it lies within the image of the masses is not the height of excellence. Nor is a victory for which the entire empire expresses its admiration and approval.

Similarly, lifting an autumn hair is hardly a sign of great strength; observing sun and moon no sign of sharp vision; and hearing the clash of thunder no sign of acute hearing. When the ancients called a fighter clever, he not only won, but did so with ease because he only conquered opponents who were easy to defeat.

For this reason, his victories did not cause celebration of his reputation for wisdom or courage. He won his battles by making no mistakes, for avoiding errors is what establishes the certainty of victory. This means conquering an opponent who is already defeated.

In other words, one who excels at warfare first establishes a position from which he cannot be defeated, and does not miss an opportunity to defeat the enemy. In this sense, the winning strategist seeks battle only after establishing the conditions for victory, whereas the losing army fights first, before seeking to assure success.

A leader of excellence cultivates the Law (Tao) and strictly adheres to method and discipline; thus it is in his power to control success.

Method
Military methodology consist of:
1. Measurement of terrain (derived from space);
2. Estimation of forces (derived from measurement of terrain);
3. Calculation of manpower (derived from estimation of forces);
4. Balancing of strength (derived from calculation of manpower); and
5. Victory (derived from all of the above).

Thus the victorious army is like a ton measured against an ounce, whereas the defeated force is like the ounce that is overwhelmed by the ton. And so the conquering force is resembles the bursting of a mountain of restrained water into a thousand-fathom valley.

5. Energy
Sun Tzu wrote: Commanding a large force is basically the same as commanding but a few. It is merely a matter of configuration, assignment and signals.

Direct (Orthodox) and Indirect (Unorthodox)

The way to ensure that your entire army can withstand the brunt of the enemy's attack unbeaten is through unorthodox and orthodox maneuvers.

Attacking the enemy like a grindstone wedged into an egg is an example of strong against weak, substantial vs. vacuous.

The usual way to engage a battle is by direct attack, although indirect tactics will also be needed in order to win. The resources of a commander gifted in applying the indirect are as infinite as Heaven and Earth and unlimited as the flowing rivers. Similar to the movement of the moon and sun, they end and then begin again, reborn like the four seasons.

The basic sounds number no more than five, and yet their combinations can produce more music than can ever be heard. Likewise, there are only five primary colors (blue, yellow, red, white, and black), but their blends create more hues than can be seen. And there are but five cardinal tastes (sour, acrid, salt, sweet, bitter), whose mixtures yield more flavors than can be tasted.

In battle, there are only two methods of attack: direct and indirect; yet in combination they enable an inexhaustible series of maneuvers. Direct and indirect flow in harmony within an endless cycle. Who could exhaust the possibilities of their union?

Force and Timing

The configuration of force (or power) is like a torrent of water pushing stones along its course.

The quality of timing (constraint) is reflected by the well-timed swoop of a falcon, which enables it to strike and overwhelm its prey. Therefore the good fighter will be focused in his attack and constrained in his timing. This configuration can be compared to releasing the trigger of a drawn crossbow.

The appearance of disorder may be simulated where there is really no disorder. In the midst of battle, your forces may appear confused and yet remain impregnable to defeat. As simulated chaos derives from control, so is (pretended) fear from courage and (feigned) weakness from strength. Order and disorder are based on numbers; courage and fear depend upon the configuration of force; and strength and weakness rely upon deployment.

Thus one who is skilled at keeping the enemy on the move forces the enemy to react to the impressions he creates. He offers something (perhaps a chariot or a few horses) that the enemy is compelled to take, keeps them moving, and then sets up an ambush.

The clever warrior relies upon the strategic effect of force without relying on his troops. This enables him to select the right men and to apply his strategy.

He utilizes strategic force to command his men like rolling logs or stones downhill. Note that wood and stone are motionless when stable, but capable of movement on a slope. Depending on their shape, it is the nature of a log or stone to remain motionless on level ground and to move when on a slope. Square-shaped forms will stop, but round ones are capable of rolling.

Thus the strategic force of a good leader is like rolling rounded boulders down a tall mountain slope. Such is the strategic configuration of force.

6. Weakness and Strength

Sun Tzu wrote: *Initiative*

In most cases, the one who occupies the field first and awaits the enemy will be prepared for battle; he who arrives later and must rush to get there will arrive exhausted. Thus does the one who is skilled at warfare impose his will upon the enemy without allowing the enemy's will to be imposed on him.

Offer the enemy the appearance of some benefit to entice him to approach of his own accord; or inflict damage to prevent the enemy from approaching.

In this manner you can tire a rested enemy, hunger him when he is well-provisioned, or force him to move when he is at rest. Take positions to which the enemy must rush to defend, and move swiftly to places where he does not expect you.

You can march a thousand li through unoccupied terrain without getting tired. To assure success, attack only positions that are undefended; to make certain your defense, secure positions that the enemy cannot attack.

Deception

Thus the skillful general attacks in such a manner that his opponent knows not where to defend; and defends in a manner that his opponent knows not where to attack. The arts of subtlety and secrecy allow us to become unseen and unheard; thus can we grasp the enemy's fate within our hands.

You may advance without resistance if you attack the enemy's weaknesses; and you can withdraw safely from pursuit if you move more quickly than the enemy. So if I choose to fight even an enemy protected behind high ramparts and deep moats, he is forced to fight because I attack targets he must save.

If I do not wish to fight, even if the lines of our encampment are only traced out on the ground, I can prevent the enemy from attacking. All I need to do is to divert his direction and confuse his movements.

Thus I learn the enemy's distribution of forces while disguising my own, enabling me to concentrate my forces where the enemy is divided. I can form a single force to attack his separated forces; if he is divided into ten groups, I attack with ten times his strength. Thus we are many where he is few. And if I am able thus to attack his small force with a superior one, he will be in dire straits

The place where we plan to engage the enemy must not be made known. This way he must prepare to defend against a possible attack at several different points; with his forces thus distributed in several directions, the forces we will face at any given point will be proportionately few.

If the enemy strengthens his front, he will weaken his rear; should he strengthen his rear, he weakens his front; if he fortifies his left, he weakens his right; and if he defends his right, he weakens the left. If he sends his men everywhere, then every side will be weak.

Numbers

Numerical weakness comes from having to prepare against possible attacks; numerical strength, from compelling our adversary to make these preparations against us.

It follows that knowing the place and the time of the coming battle enables one to cover a thousand li and concentrate to engage in battle. But if one knows neither the time nor place of battle, then the left flank will be unable to help the right, the right equally unable to support the left, the front unable to relieve the rear, nor the rear to support the front. How much more so when the most distant are separated by some tens of li and even the nearest by several li!

According to my estimate, the soldiers of Yueh exceed our own in number, but what advantage will that bring them in attaining victory? Thus I say that victory can be achieved. For although the enemy be stronger in numbers, we can prevent them from fighting.

Study the enemy to discover their plans and the likelihood of their success. Stimulate them to identify the pattern of their movement and inactivity. Force them to reveal themselves to uncover their vulnerabilities. Carefully compare their army with your own to learn where they are strong or weak.

The ultimate of military deployment is to conceal your troops without a visible form. This will keep even the most clever spy from observing and making plans against you.

Tactics

The manner of producing victory from the enemy's own tactics, this is what the masses cannot understand. All men can see the tactics by which I conquer, but none can see the strategy from which my victory is gained.

Do not repeat the tactics which have gained a victory, but let your methods be influenced by the infinite variety of circumstances.

Now tactics are like water; for water's natural course flows away from high places and rushes downward. Thus in war, the way is to avoid what is strong and to strike at what is weak. As water configures its course according to the terrain over which it flows, the army controls the course of victory in accordance with the opposition. Just as water keeps no constant shape, in warfare there are no constant conditions.

One who is able to modify his tactics in relation to his opponent, and thereby succeed in winning, is known as a spiritual leader.

So it is that no one of the five elements (water, fire, wood, metal, earth) will dominate; nor do the four seasons resist their transitions into one another. Days grow long and short, and the moon continues to wax and wane.

7. Maneuvering

Sun Tzu wrote: In war, the general commonly receives his commands from the ruler. He begins by assembling the army and organizing the masses*, then goes on to confront the en-

---

* Sun Tzu is referring to the professional (trained) army and the masses of untrained conscripts.

emy and set up his forces. Next comes the most difficult step, that of tactical maneuvering, which is the process of turning the devious route into a direct path, and adversity into gain.

The art of deviation here is to entice the enemy (with the illusion of gain) into taking a long and circuitous route, so that even though starting after them, you manage to arrive before them. This maneuver is accomplished through the tactics of the circuitous and the direct.

Logistics

Combat with an army is advantageous; with undisciplined masses it is dangerous. If you march a fully equipped army order to seize an advantage, you will arrive too late. But if you send forth a smaller force, you will lose your baggage and equipment.

If you march a hundred li by setting aside your armor and rushing forward, day and night without stopping, advancing twice the usual distance at a stretch, covering a hundred li in order to gain an advantage, the leaders of all your (three) divisions will be captured: the stronger men will arrive first, the exhausted ones will fall behind, and only one in ten of your army will reach its destination.

If you march fifty li in order to outmaneuver the enemy, you will lose the leader of your first division, and only half your force will reach the goal.

If you march thirty li in the same manner, two-thirds of your army will arrive.

Clearly, an army without its baggage and heavy equipment is lost; without provisions it is lost; without supply stores it is lost.

Familiarity

We cannot enter into alliances until we are acquainted with the designs of our neighbors.

We cannot lead an army unless we are familiar with the lay of the land – its mountains and forests, its gorges and precipices, its marshes and swamps. We cannot turn profit from advantages of terrain without making use of local guides.

In war we need to practice deceit, advantageous maneuvers and flexibility. Your decision to concentrate or divide your troops must be decided by circumstances.

Let your speed be that of the wind, your order that of the forest; your raids and plundering like fire; your defense as immovable as a mountain. Let your plans be as dark and impenetrable as the night; your movement like a thunderbolt.

When you plunder, divide the wealth amongst your men; when you capture new territory, divide the profits among the soldiers.

When you move, control the strategic balance of power. He who first masters the tactics of the circuitous and the direct will win. Such is the art and strategy of military combat.

The Book of Army Management states:
On the field of battle, the spoken word does not carry far enough, for which reason gongs and drums were created. Nor can they clearly see one another, for which reason they made banners and flags. Gongs and drums, banners and flags are the means by which to focus men's ears and eyes. In forming a single united body, brave men will not have to advance alone, nor will cowards be able to retreat alone. This is the art of handling large masses of men.

In night fighting, then, use signal-fires and drums, and when fighting by day, use flags and banners to influence the ears and eyes of your army.

Mental Discipline
An entire army may be robbed of its spirit, the commander's mind overtaken. The soldier's spirit is strongest in the morning; by noon it has begun to weaken; and in the eve-

ning his mind turns toward returning to camp. A clever general therefore avoids an enemy whose spirit is strong, but instead attacks when the opposition is sluggish and longs to regain its campgrounds. This is the way to manipulate the enemy's spirits.

Disciplined and calm, he awaits disorder amongst the enemy: this is how to remain self-possessed.

To be near while awaiting those who are distant; to be rested while awaiting the fatigued; to be well-fed while the enemy is hungry: this how to manipulate strength.

To refrain from intercepting an enemy whose flags are in perfect order; to refrain from attacking a confident and well-ordered army: this how to manipulate circumstances.

Military wisdom dictates against advancing uphill against the enemy, or to oppose him when he descends. Do not pursue an enemy who pretends to flee; do not attack animated and well-ordered troops; so not swallow bait offered by the enemy; and do not interfere with an army that is returning home.

When you surround an army, always leave them a way out.
1. Do not press a desperate foe with too much force.
2. Such is the art of warfare.

8. Tactical Variations

Sun Tzu wrote: Here is the normal procedure for conducting military matters: the general receives his commands from the ruler, unifies the army, and concentrates his forces.

His principles are:
1. Do not encamp on dangerous (entrapping) terrain.
2. On terrain where major roads intersect, unite with your allies.
3. Do not linger on isolated terrain.

4. Use strategic planning for encircled terrain.
5. On fatal (desperate) terrain you must fight.
6. There are roads that must not be followed.
7. There are armies that must be not attacked
8. There are fortified cities that must not be sieged.
9. There is terrain that must not be contested.
10. There are commands from the ruler that must not be
    followed.

Thus the general who thoroughly understands the advantages of tactical variations knows how to use his troops. Whereas the general who does not understand these, even if familiar with the configuration of terrain, will not be able to profit from his knowledge.

It follows that one who is well-versed in the art of varying his plans, and is acquainted with the Five Advantages [but does not know the techniques of the nine changes*], will be unable to lead and control his men.

The wise leader calculates the sum of gain and loss. If he perceives an advantage in the midst of difficulty, or a potential downside within an apparent advantage, his judgment can be trusted.

To subdue the enemy (feudal lords), inflict damage on them, keep them constantly engaged by making trouble for them, and lure them here and there for any number of imaginary profits.

The art of war teaches us to rely not on the likelihood of the enemy's not coming, but on our readiness to met him; not on the chance that he will not attack, but on the fact that we have made our position unassailable.

There are five dangerous character faults which may affect a general:

---

* This reference to "nine changes" has confused scholars and translators for centuries; we can only surmise that they refer to the (10) points listed above.

1. Recklessness, which leads to destruction;
2. Cowardice, which leads to capture;
3. A quick temper, which can be provoked by insults;
4. Obsession with honor, which is sensitive to shame;
5. Over-solicitude for his men, which exposes him to worry and trouble.

These are the five dangerous traits that can lead to disaster in a war. The destruction of an army and the leader's death invariably stems from these faults, and so one must be wary of them.

9. On the March

Sun Tzu wrote: *We come now to the question of encamping the army, and observing signs of the enemy.*

Terrain

To cross mountains, seek valleys facing the sun and pass quickly. Camp in high places. If the enemy is encamped in the heights, do not climb up to battle them. This is mountain warfare.

After crossing a river, move far away from it. If the enemy is crossing a river in your direction, do not attack them in midstream. First let half the army cross, then attack.

If you want to engage in battle, do not approach the enemy near a river which he has to cross. Rather, seek higher ground than that of the enemy facing the sun. Do not move against the current to meet the enemy. This is river warfare.

When crossing salt marshes and wetlands, focus on getting across as quickly as possible, and then moving away from them without lingering. If you must fight in a salt marsh, seek an area close to water and grass, with trees to your back. This is wetlands warfare.

On dry and level terrain, take an easily accessible position with elevated ground to your right and rear, so that the danger lies ahead and safety to the rear. This is flat terrain warfare.

These advantageous forms of military deployment enabled the Yellow Emperor to vanquish the four emperors.

Yin and Yang

All armies prefer high ground to low, and sunny (yang) places to the dark (yin). Camp close to grass and water to avoid the hundred illnesses, and you will be assured of victory.

Where there are hills and embankments, occupy the sunny side with the slope to your right rear. This will benefit your soldiers and exploit the natural assistance of the ground.

When heavy rains occur upstream, foam appears upon the water. If you wish to cross the river, wait until it subsides.

Move quickly away from terrain with such deadly configurations as precipitous cliffs separated by swift torrents, deep gorges, confined areas, tangled thickets, quagmires and crevasses. These should not be approached.

By avoiding such places, we may cause the enemy to approach them. Thus when we face the enemy, these terrains will be at his rear.

Signs

If within the neighborhood of your camp there are any hills, wetlands with tall reeds and grass, forests with thick undergrowth, you should carefully search them, for these are places where ambushes and spies may be hiding.

1. If an enemy is nearby and remains quiet, he is relying on the natural strength of his position.
2. If he stays away and tries to provoke you into battle, he wants you to advance because he holds an advantage of terrain.

3. If his position offers easy access, he is tendering a bait. Movement in the trees shows that the enemy is advancing. The appearance of obstacles in the thick grass means that the enemy wants to make us suspicious. If birds take flight, the enemy is setting an ambush. If the animals display fear, a sudden attack is on the way.

4. When dust rises in a high column, chariots are advancing; when the dust is low and spread over a wide area, the infantry approaches. If the dust branches out in different directions, they have gone to gather firewood. Smaller clouds of dust that comes and goes are signs that the enemy's army is encamping.

5. They who speak softly but increase their preparations are about to advance. They who speak with belligerence while advancing quickly will retreat.

6. When their light chariots deploy to the sides, they are preparing for battle.

7. When they offer proposals without prior conditions, expect a devious plot.

8. When there is much running about and the soldiers are formed into rank, the critical moment has come.

9. When some enemy forces are seen advancing and some retreating, it is a lure.

10. If they stand about leaning on their spears, they are hungry.

11. If those who draw water are the first to drink, then they are thirsty.

12. If they see an advantage to be gained and make no effort to secure it, they are exhausted.

13. Wherever birds may gather is an empty place.

14. If the enemy cries out at night, they are afraid.

15. If there is a disturbance in the camp, their leader is weak.

16. If their flags and pennants are moved at random, they are confused.

17. If the officers are angry, then the men are weary.

18. If they kill their horses for food, they lack grain.
19. If they hang up their cooking-pots over the campfires and do not return to camp, it means that they are desperate and determined to fight to the death.
20. When the men whisper together in small groups, the leader has lost the masses.
21. When he rewards his men too frequently, the enemy is at the end of his resources; too many punishments betray a dire distress. When he begins with bluster and then becomes fearful, he shows a supreme lack of intelligence.
22. When envoys are sent with offerings, the enemy wishes for a truce.
23. If the enemy's troops march up angrily and remain facing yours for a prolonged without either joining battle or taking themselves off again, you should investigate the reason very carefully.

Leadership
Even if your troops number no more than the enemy's, that is sufficient, for he cannot advance. It is enough to concentrate your available strength, keep a close watch on the enemy, and wait for the right moment to attack.

Only if you exercise no forethought and underestimate your opponent are you likely to be captured by them.

If you punish soldiers before they have grown attached to you, they will not be submissive; in this case, they will be difficult to assign. If you fail to enforce punishments after the soldiers have become attached to you, they will still be useless.

Therefore, soldiers must be treated first with humanity, but kept under control with iron discipline. This is a certain road to victory.

If commands are consistently enforced, the army will be well-disciplined; if not, its discipline will be poor.

A general who shows confidence in his men but always insists on his orders being obeyed will establish an advantageous mutual relationship.

10. Terrain

Sun Tzu wrote: *Different Terrains*

There are six major kinds (configurations) of terrain:

1. Accessible: This is terrain into which our forces and those of the enemy can both advance. We occupy the high ground and sunny side, and maintain our supply routes in order to battle with advantage.

2. Entangled: This is terrain that is easy to enter but from which it will be difficult to withdraw. If the enemy is unprepared here, we may be able to go forth and defeat them. But if the enemy is prepared and we advance without winning, we will be unable to withdraw and may suffer disadvantages.

3. Temporizing: This is terrain where it is disadvantageous for us or the enemy to advance. So even if the enemy tries to bait us with the appearance of some sort of gain, instead of advancing we retreat. If part of their army comes forth, we may strike at them with advantage.

4. Constricted: This is terrain with narrow passageways. If we occupy it before the enemy we must establish strongholds throughout to await them. If they occupy it before us and are thoroughly deployed, we will not follow them inside. But if the enemy positions are not well manned, we may follow them.

5. Precipitous: This is terrain where we need to occupy the sunlit heights to await the enemy. But if they occupy these positions first, withdraw and do not follow.

6. Distant: This is terrain that lies at a great distance from the enemy. Given equal strategic force, it will be difficult to provoke conflict or to gain an advantage.

Conditions for Failure

There are six conditions that can cause an army to fail. They arise not from natural causes (Heaven and Earth), but from the general's mistakes.

1. Flight: All other conditions being equal, attacking a force ten times your size will result in flight.
2. Insubordination: If the soldiers are strong and their officers weak, this will result in their disobeying orders.
3. Collapse: If the officers are strong and their troops weak, this will result in their collapse.
4. Ruin: If the higher officers are angry and insubordinate, and engage the enemy out of resentment before the commander-in-chief has assessed their capability, the result is ruin.
5. Chaos: If the general is weak and lacks authority, unclear in his commands and leadership, fails to assign clear responsibilities to his officers and men, and deployment is handled in a haphazard manner, the result is utter chaos.
6. Rout: If a general, unable to estimate the enemy's strength, allows a small force to engage a larger one, or launches a weak detachment against a powerful one, and neglects to place picked soldiers in the front rank, the result will be a rout.

These six ways of courting defeat* must be carefully examined by the general in charge.

Generalship

The natural configurations of terrain are the soldier's ally; ability to assess the adversary, control the forces of victory, and shrewdly calculate the difficulties, dangers and distances of terrain are the test of a great general.

---

* The "Tao of defeat."

The one who knows these things and applies them to battle will surely be victorious. The one who does not know or practice them will surely be defeated.

If victory is assured, you must engage even if the ruler has forbidden it; if victory is unlikely, you may avoid fighting even if the ruler has told you to engage.

So it is that the general who advances without seeking fame and retreats without fearing disgrace, whose only thought is to protect his country and do good service for his sovereign, is the jewel of the kingdom.

If you regard your soldiers as your children, they will follow you into the deepest valleys; look upon them as your own beloved sons, and they will stand by you even unto death.

If, however, you are indulgent but unable to exercise your authority; kind-hearted but unable to enforce your commands; and incapable of quelling disorder, then your soldiers are like to spoiled children and are useless.

If you know that your own men are in a condition to attack, but are unaware that the enemy is not vulnerable to attack, you have gone only halfway to victory. If you know that the enemy is vulnerable to attack, but are unaware that your own men are unable to attack, you have gone only halfway towards victory.

If you know that the enemy is vulnerable to attack, and also know that your men are able to attack, but are unaware that the terrain is unsuitable for attack, you have still gone only halfway towards victory.

Thus the one who is experienced in these matters will never be bewildered when he goes forth, never at a loss when he initiates an action.

So it is said: If you know the enemy and know yourself, your victory will not be in doubt; if you know Heaven and Earth, your victory can be complete.

11. The Nine Varieties of Terrain

Sun Tzu wrote: The art of war recognizes nine varieties of terrain: dispersive, facile, contentious, open, focal, serious, difficult, encircled, and desperate.

1. Dispersive: This is where a chief is fighting within his own territory.
2. Light (frontier): This is where one enters hostile territory, but not very deeply.
3. Contentious: This is where it would be advantageous for either side to occupy.
4. Open: This is where either side can go.
5. Focal: This is contiguous to three other states, such that the first to control it may gain the support of these neighboring states.
6. Serious: This is where one has penetrated deeply into enemy territory having left many fortified cities to the rear.
7. Difficult: This is where there are mountain forests, rugged steeps, wetlands, and terrain that is difficult to navigate.
8. Encircled: This is where access is restricted and from where return is difficult and indirect, enabling the enemy to strike with a small body of men.
9. Desperate (fatal): This is where one must fight with desperation to avoid being destroyed.

For these reasons:

1. On dispersive ground, do not engage the enemy;
2. On light ground, do not stop;
3. On contentious ground, do not attack;
4. On open ground, do not permit your men to get isolated from one another;
5. On focal ground, unite with the nearby rulers;
6. On serious ground, plunder for supplies;
7. On difficult ground, move quickly through;
8. On encircled ground, resort to strategy;
9. On desperate ground, fight with all your strength.

Old Lessons

Back in ancient times, those who were called skillful leaders knew how to drive a wedge between the enemy's front and rear; to prevent trust and co-operation between his large and small divisions; to keep the better troops from rescuing the lesser, the officers from rallying their men.

When the enemy's men were united, they knew how to sow disorder amongst them. They moved when it was to their advantage and stopped when not to their advantage.

If asked how to cope with a great host of the enemy in orderly array and on the point of marching to the attack, I would say: "Begin by seizing something your opponent values, for he will then listen to you."

Speed is the essence of war, for it enables you to take advantage of the enemy's absence, travel unexpected routes, and attack unguarded positions.

General Principles

These are the principles to be observed by an invading force:

The further you penetrate into foreign territory, the greater the solidarity of your troops, and thus the defenders will not be able to stand against you.

Make forays in fertile terrain to feed your army.

Attend to the well-being of your men and do not over-exert them. Concentrate your energy and maintain your strength. Keep your army continually on the move, and devise unfathomable plans.

Put your soldiers into positions from which there is no escape, and they will fight to the death without retreating. Facing death, officers and men will give all their strength. Soldiers in desperate straits will lose their sense of fear and will stand firm. When they have no alternative they will fight hard.

Thus, even without instruction they will be ready; without waiting to be asked they will cooperate; without restrictions they will be loyal; without orders they can be trusted. Prohibit superstition, and eliminate rumors and doubt, so that nothing but death itself be feared.

If our soldiers have little material wealth, it is not because they distain riches; if their lives are not very long, it is not because they dislike longevity.

On the day they are ordered out to battle, your soldiers' tears will soak their sleeves and roll down their cheeks. But once they face a hopeless fight they will display the courage of the immortals Zhuan Zhu and Cao Kuei.

The skillful tactician may be compared to the shuai-jan, a snake that is found on Mount Ch'ang. Strike at its head and the tail will respond; strike at the tail and the head will respond; strike at its middle and both head and tail will attack.

Asked if an army can be made to imitate the shuai-jan, I would answer that it can. Consider that the men of Wu and those of Yueh are enemies; but if they were crossing a river in the same boat and were caught in a storm, they would support each other just as the left hand helps the right.

Thus tethering the horses and burying the chariot wheels are not enough to keep the soldiers from fleeing. The principle of army management is to set up a standard of courage that applies to all. This is done through the appropriate use of hard and soft patterns of terrain. Thus the skillful general leads his army in the same way as leading a single man by the hand.

A general needs to be calm and quiet to ensure secrecy, upright and disciplined to maintain order while keeping officers and soldiers ignorant of his plans. He varies his direction and changes his strategies to avoid their being perceived. He modifies his position and takes indirect routes to prevent the enemy from anticipating his plans.

At the critical moment, it will seem to the troops as if they had climbed up a height and then kicked away the ladder. The general accompanies his men deep into hostile territory before he shows his hand. He burns his boats and breaks his cooking-pots; like a shepherd driving a flock of sheep, he pushes his men this way and that without any of them knowing their destination.

Assembling his forces and putting them in danger. This is the general's responsibility.

Invasion Principles
The different measures suited to the nine varieties of ground; the expediency of aggressive or defensive tactics; and the fundamental laws of human nature: these are things that must certainly be studied.

In hostile territory, the general principle is that penetrating deeply brings cohesion among your troops, but shallow penetration inclines them to disperse.
1. When you leave your country behind and take your army into enemy territory, you are on critical terrain.
2. When all four sides are open, this is focal terrain.
3. When you penetrate deeply into a country, it is serious terrain.
4. When you penetrate only a little way, it is light terrain.
5. When you have enemy strongholds behind you and narrow ways in front, it is encircled terrain.
6. When there is no else to go, it is desperate terrain.

Therefore:
1. On dispersive terrain I inspire my men with unity of purpose.
2. On light terrain I keep them close together.
3. On contentious terrain I speed up the rear.
4. On open terrain I see to my defense.
5. On focal terrain I consolidate alliances.
6. On serious terrain I ensure a continuous stream of supplies.
7. On difficult terrain I advance quickly.

8. On encircled terrain I block any openings.
9. On desperate terrain I let them know that our lives cannot be saved.

For it is the nature of a soldier to defend when surrounded,
to fight with passion when there is no alternative, and to
follow orders when the situation dictates.

One cannot enter into alliances with neighboring princes
until he knows their plans. One who is unfamiliar with the
lay of the land: the mountains and forests, pitfalls and precipices, wetlands and marshes cannot lead. One cannot profit from natural advantages without the use of local guides.
One who is unaware of any one of these principles cannot
command the army of a true ruler.

When a true leader attacks a powerful state, the enemy's
forces are unable to assemble. His generalship shows itself
in preventing the concentration of the enemy's forces. He
overawes his opponents, keeping their allies from joining
against him.

For this reason he does not seek just any and all available
alliances or support the power of other states. He carries
out his own designs, leaving his antagonists in awe. Thus
he is able to capture their cities and overthrow their rulers.

Bestow rewards without regard to rules of law, and impose orders beyond previous arrangements. Command
your forces as if commanding but a single man. Issue orders
without explaining their purpose, and offer the possibility
of profit without telling them of the dangers.

Put your army into deadly situations and it will survive;
lead them into desperate terrain and they will endure. For
it is only after they have been exposed to danger that they
will be capable of snatching victory from defeat.

The key to successful military operations is in learning
and accommodating oneself to their intentions. If you concentrate your efforts toward the enemy, you can strike from

a distance of a thousand li and kill their general. This is called "being skillful in a clever and creative manner."

This is why, on the very first day of your command, you block the frontier passes, destroy official tallies, and stop the passage of all emissaries. Address the members of the council firmly to control the situation and gain their support.

If the enemy opens a door, you must rush in.

Attack that which he values most. Maintain a flexible timetable for battles; assess and react to the enemy so as to plan your strategy.

At first, exhibit the coyness of a shy maiden, until the enemy provides an opening; then run as swiftly as a hare. The enemy will not be able to understand you or oppose you.

12. Attack by Fire

Sun Tzu wrote: There are five ways to attack with fire. The first is to burn soldiers in their camp; the second is to burn the enemy's provisions; the third is to burn their supply trains; the fourth is to burn their armories; and the fifth is to burn their transportation facilities.

Carrying out an incendiary attack necessitates the proper means and conditions. Incendiary materials should be prepared and kept in readiness.

Certain seasons are more opportune to initiate attack by fire, and certain days are ideal to ignite them. These are when the weather is driest, and the best days are when the moon is in the constellations of the Sieve (chi), Wall (pi), Wing (i) or Cross-bar (chen), for these four are the days of rising wind.

In attacking with fire, you must be prepared for five possible developments:
1. When a fire starts inside the enemy's camp, respond at once with an attack from outside.

2. If a fire starts but the enemy's soldiers remain quiet, wait without attacking.

3. When the flames have reached their highest point, attack if possible; otherwise remain where you are.

4. If it an assault with fire can be made from outside, do not wait for it to happen on the inside, but deliver your attack at the most favorable moment.

5. Start the fire from the upwind side; do not attack from downwind.

A wind that rises in the daytime will last, but a night breeze will soon fail.

An army should be aware of the five possible developments of fire so they can defend against them during the appropriate seasons and days. The ability to use fire as an aid to the attack is enlightened, and using water as an aid to the attack adds strength. Water can be used to isolate the enemy, but not to capture their supplies.

Winning battles and attacking successfully without exploiting gains is wasteful and stagnating. Thus it is said that the enlightened general lays his plans well ahead, and the good general cultivates his success.

Move only if you perceive an advantage; do not use your troops unless there is something to be gained; do not fight unless the position is critical. A ruler should not send forth his troops merely out of personal anger; and a general should not engage in battle out of frustration. If it is to your advantage, make a forward move; if not, stay where you are.

Anger can give way to joy, annoyance to contentment. But a kingdom once destroyed cannot be restored; nor can the dead be brought back to life.

Hence the enlightened ruler is cautious, and the good general respectful. This is the way to keep the state at peace and the army intact.

13. Using Spies

Sun Tzu wrote: Raising an army of a hundred thousand men and sending them a thousand li will wreak a heavy toll upon the people and drain the resources of the State. The daily expense will run a thousand pieces of gold. Seven hundred thousand families will be interrupted from their farm work on both sider of the border, and men will fall exhausted along the road.

Hostile armies will confront one another for years, striving for a victory to be decided in a single day. If a general begrudges the expense a hundred pieces of gold and thus remains in ignorance of the enemy's condition, he is devoid of humanity. Such a man is not a general for the people, no help to his ruler, nor a master of victory.

Wise rulers and competent generals are able to strike, conquer, and achieve results beyond the ordinary due to advance knowledge. This cannot be gained from the supernatural, inferred from experience, nor deduced by calculation. Information on the enemy's dispositions can only be obtained from men who have this knowledge.

Five different types of spies may be employed, including native, internal, converted, expendable, and surviving:
1. Native: people from among the enemy;
2. Internal: enemy officials;
3. Converted: double-agents converted from the enemy's agents
4. Expendable: one's own agents who are fed false information to pass on to the enemy;
5. Surviving: those who return with their information.

Spying
When these spies are put to work at the same time, this "spiritual manipulation" is the ruler's most precious resource. Thus it is that of all military affairs, no relationship

is more intimate than that with spies; none more liberally rewarded, and no arrangements more secret.

Spies cannot be usefully employed without the wisdom of a sage, and cannot be managed without generosity, benevolence and humanity. Without subtlety and perception, the validity of their reports cannot be properly evaluated and put to use.

This is subtle, subtle work; for spies are used in every kind of situation. If a secret piece of news becomes exposed before its time, the spy and all whom he informed must be put to death.

Whether your objective is to attack an army, storm a city, or assassinate individual people, you must first learn the names of the commander, his assistants, staff, bodyguards and sentries. You need to send your spies to obtain this information.

Enemy Spies

Seek out the opposition's spies; tempt them with bribes, and then provide instruction and retraining. This is how double agents are obtained and employed. Through their information we can then recruit and employ both local and internal spies. And through the information they provide, our expendable spies can misinform the enemy with false details; and our surviving spies can also be employed as needed.

The purpose of spying in its five varieties is knowledge of the enemy, which can only be derived from converted spies. For this reason, these double agents must be treated with the greatest generosity.

In former times, the rise of the Yin dynasty was enabled by I Chih, who had served under the Hsia. Likewise, the rise of the Chou dynasty was due to Lu Ya, who served under the Yin.

Thus it is that the enlightened ruler and the wise general who employ highly intelligent spies are able to achieve great results. Spies are a vital element in warfare, because the military depend on them to make their moves.

PART TWO

Sun Tzu for Women

*"...supreme excellence consists in breaking
the enemy's resistance without fighting."* – Sun Tzu

The Art of War was written for the military leaders of the time, virtually all of whom were men. Fortunately, these strategies and tactics are universal, not gender-specific: they can be (and have been) successfully used by clever women and men all over the world.

Women and The Art of War addresses ways in which women can use Sun Tzu's principles to help them succeed in a wide range of competitive environments. The differences in scope and audience lie not so much within the teachings of Sun Tzu, but rather in the ways that women can embrace and apply them to their own advantage.

1. Planning

*"Warfare is of vital importance to the state, the basis of life
and death, the way to survival or extinction.
"Therefore, it is essential to structure it according to the pros
and cons of the five constant factors."* – Sun Tzu

This chapter introduces the five main pillars on which *The Art of War* is based:

1. Integrity
2. Conditions
3. Obstacles
4. Leadership/credibility
5. Management/policy

The topics introduced here, and their significance to women, will reappear in greater detail and different contexts throughout this book.

Integrity

*"Righteousness (integrity) is the force
that underlies creation." – Sun Tzu*

Sun Tzu emphasized integrity as an underlying force, or moral checkpoint, to the warrior's ambitions. Why is this important to women?

Several reasons, the first of which may be to question the status quo of blind ambition established and widely accepted largely by the male community. Another potential issue is that the newest kid on the block—i.e., the visible minority (including women)—is likely to attract more than her fair share of attention. Thus small errors of judgment and, of course, indiscretions, tend to be magnified and remembered.

Basically, the message is to do the right thing—and avoid doing the wrong thing. This begins with asking yourself why you want to do something, and what is the likely result if you succeed?

Example:
Why do I want to be selected as the project manager?
a. To grow as a professional?
b. To motivate the team?
c. To get a promotion?

d. To get a raise?
e. To gain recognition as a woman?
f. To win a bet?
g. To keep someone else from getting it?

Personal growth and motivating yourself and others (a and b) are, in most cases, healthy and inspiring. In combination, they suggest an ideal path for women (men too, when they slow down long enough to take stock).

Seeking promotions and raises (c and d) can be worthwhile as long as you consider the cost of succeeding. Are you taking this track just because the guys are doing it, or is it something you really want for yourself? Are you prepared to rise to the challenge of elevated expectations in exchange for more money and status? If so, go for it and good luck! But at least think about the pros and cons, and consider whether the new responsibilities are your cup of tea.

A goal of recognition as a woman is a tricky one and may prove to be a bad choice even with the best of intentions. While gaining recognition is admittedly important, this motive has the potential to lead you astray. If you really want the job, promotion, etc., there's nothing wrong with using the recognition factor to add fuel to your enthusiasm. Beyond that, it may be no better than the latter two incentives.

You have surely recognized that the final pair (f and g) aren't worth your effort. For one thing, they encourage the futility of trying to get something you may not be suited for, much less want. Even more importantly, they are negative pursuits. Since you may have to work harder than most men to prove yourself, why waste your time and effort on a dead-end path?

There's an old saying that people would throw more things away if they weren't afraid that someone else might pick them up. Don't allow such negative motivations to influence your decisions and aspirations.

It isn't selfish to do what works for you.

Being true to yourself puts you in a more comfortable position and frame of mind to do right by those around you. Treating colleagues fairly, lending an occasional helping hand, and being known as a reliable person are likely to benefit you as well. Granted, there may be a few individuals who won't like you no matter what you do, but that is more often their problem than yours.

We're not suggesting that you sacrifice your interests to allow your colleagues to succeed. That's silly and unnecessary. Rather, maintain your code of ethics, do what feels right and seems to work for you, and establish an equilibrium between career goals and your sense of self.

Ambition for its own sake has long been associated with the male of our species, while women are known to be more nurturing. Well, don't be mislead into thinking that ambition is anti-feminine, or that you'll have to emulate a bunch of "gentlemanly" characteristics in order to succeed—especially not when you can pursue a professional and healthy balance as a successful woman! We will pursue these ideas in greater depth throughout the following chapters.

Synopsis
If you are true to yourself, fair to others, and consistent, the rest will follow.

Taking Advantage
"When it is advantageous, move; when
not advantageous, stop." – Sun Tzu

The best time to advance, retrench, or step back depends on what is happening around you. Efficiency relates to your approach to the conditions at hand; when you are swimming against an uneven current, you can't afford to waste your energy.

The Art of War teaches that timely use of defensive measures can be as aggressive as attack. Blindly rushing forward exposes you to obvious dangers and risks damaging potential allies and resources; preparation, a strong defense, and recognizing when to use them are among the keys to a successful campaign. Sun Tzu spoke of terrain, seasons and resources when determining tactics. These conditions are applied to useful tips and strategies for women in the later chapters of this book.

The Master also reminds us of the universal balance represented by the Chinese terms, yin and yang. In this context, it is particularly important to understand their relevance to women.

Yin, which represents the female and the earth, is traditionally associated with turmoil, darkness, cold, conservation and substance; yang – the male side – stands for heaven, light, heat, destruction and disintegration.

Yin is soft; yang is hard. Yin is moon; yang is sun.
Yin is mountain; yang is river. Yin is intuition; yang is logic.

We hasten to point out that these are intended as complementary universal forces, not as definitions of the differences between women and men. All living creatures, including humans, share these qualities to a greater or lesser degree. According to traditional Chinese medicine, a person's health is optimum only when their yin and yang are in balance; indeed, illness is believed to be the result of one or more imbalances.

What does this mean to women?

The message is threefold:
1. Despite so-called "conventional wisdom," neither gender has a monopoly on thought or behavior patterns.
   But to shore up, tone down, or otherwise modify your

natural strengths and weaknesses, you need to first
identify and acknowledge them.

2. In all things, you are best served by a healthy and productive balance between contrasting qualities. Since
   women tend to be more grounded than their male counterparts, they can more easily learn to recognize when to
   emphasize a certain quality and how much of it to use.

3. Wisdom, strength, courage, purpose, skill, analysis, discipline, and all the other qualities that may be needed
   to succeed in a variety of circumstances are as available
   to gals as to guys. Applying them in proportion to the
   need at hand may provide you with a strategic advantage over those with heavy hands or fragile fingers.

Western societies tend to focus upon opposites, whereas
Sun Tzu and traditional Chinese philosophy view differences as degrees (yin/yang) of the same thing. If this seems
counterintuitive at first, it will grow familiar as you dig
deeper into the pages and ideas that follow.

Synopsis

Come to terms with your assets and liabilities, and recognize
when to move forward, retrench, or step back. Innovators
do not go against the flow; they create their own direction.

Embracing Obstacles

*"Earth embraces far or near, difficult or accessible, open or
restricted, dangerous or safe conditions." – Sun Tzu*

Women know that opportunities come with a price tag.
Whether you pay up front or later on, nothing worth having is for free. That's why career and other significant decisions must be carefully weighed and evaluated: probable
cost vs. potential gain.

What kinds of cost?

First, take the time and effort to learn and work toward an objective. This can entail study (formal and ad hoc), periods of long hours and hard work (apprenticeship), sacrifice and patience.

Women are often expected to juggle family obligations and career ambitions like magicians. Are these your expectations of yourself?

More subtle, but no less real are the interpersonal relationships you form along the way. And here, more than in any other facet of their careers, is where women are faced with greater obstacles – call them complications – than their male counterparts.

Recognizing

It is no secret to women that sexism is an ugly social and career obstacle to women here on planet Earth. In the US and most developed nations, sexual discrimination is illegal but in full regalia; in many other places it remains the status quo.

Back in the 17th century, poet George Herbert wrote: "Words are women, deeds are men" (this from a man of letters). Two centuries later, Mathew Arnold wrote: "With women the heart argues, not the mind." Haven't attitudes like these finally disappeared? Not entirely. Given the sensitivity of our society and working environments today, sexist sentiments are voiced far less freely than ever before. But regardless of what is said or not said, women continue to battle for equal pay, equal opportunity for advancement, and the sort of assignments that will challenge them and allow them to prove their value in the workplace. The reasons for this are many, including outmoded ideas of what women can offer. What many men have yet to realize is that women like Catherine the Great, Joan of Arc, Margaret Mead, Golda Meier, Marie Curie, Indira Gandhi, Chien-Shiung Wu, Mother Teresa and Benazir Bhutto (to skim the surface) are not the exception that proves the rule.

In the 20th century, Marianne Williamson—author, lecturer and minister—wrote in *A Return To Love: Reflections on the Principles of A Course in Miracles*: "Our deepest fear is not that we are inadequate. Our deepest fear is that we are powerful beyond measure.... We ask ourselves, 'Who am I to be brilliant, gorgeous, talented, fabulous?' Actually, who are you *not* to be?"

The hope is that, in these times, all women realize that the perceptions that have followed them for so long are wrong. Actual liabilities, such as limited experience and poor personal, analytical and communications skills, apply to men and women equally and can be overcome—or at least improved—regardless of gender. Liabilities of perception, however, are harder to shake off, and continue to trap women in dead ends. While women are indeed powerful beyond measure, it's wise to be prepared for the fact that some battles continue to be tough—and some will be tougher than others.

Back in Sun Tzu's time, women were known to be skillful on horseback. Their smaller and lighter stature enabled their mounts to carry them faster and for greater distances, but restricted their ability to wield heavy swords and weapons. Technology has erased this disadvantage, and women are now recognized as being at least equal to men in marksmanship, piloting planes and navigating ships. Nevertheless, the military remains a bastion of male dominance by sticking to its perception of the gentle sex. And this is just one example of a male-dominated profession.

As you narrow in on your chosen career, examine carefully the lay of the land. Achieving success in most fields is tough enough; the addition of closed doors, glass ceilings and, perhaps, hostile competition, should also be considered.

Experience has shown sexism to be stronger in certain professions and organizations than others. Depending on your tolerance for confronting obstacles, this can be a legitimate concern in deciding where to send your resume. Just as you'd go slow and check for potholes when jogging on unfamiliar ground, you'll want to know what shadows may be lurking along the path of your career objectives.

You are also advised to take stock of your personal strengths and weaknesses (lesser strengths) to assure that the former will be useful and the latter will not hold you back before you've had a chance to shore them up.

Overcoming

What of embracing obstacles? Generally, there are two types of obstructions women are likely to encounter in the workplace: predictable and random. The majority (often imposed by other women as well as by men), are well known and documented. You need to be prepared for these, as we shall detail later in this book. The key is to be realistic in your expectations: observe what is, not what you might like it to be.

No, we're not suggesting that you should accept less than equal treatment, only to be aware that unpleasant situations may arise from time to time, and you will need a way of dealing with them in harmony with your personality and comfort zone. As for random occurrences that catch you by surprise, the confidence you develop from handling the obvious problems will equip you to ad lib more effectively.

The initial steps lie in anticipating and recognizing gender-based attitudes, as we have been discussing. The fun part is when you get to use negative perceptions about your abilities to your advantage. Envision a steep incline: advantage or disadvantage? Well, now, doesn't that depend whether you are on the top looking down, or on the bottom looking up?

In all circumstances, be yourself. Whatever your personality, ambitions, background or status, bear in mind that

you are an individual and a woman, not just a worker/
competitor.

Dress for the occasion and to suit your style, preferably
in a feminine (but not provocative) manner: if most of your
colleagues are wearing suits or business casual, there are
ways to express your individuality while staying in synch.

Develop professional relationships with a few men and
women whose character you admire and respect.

Let males perform common courtesies for you, like opening doors, when they are so inclined.

Be friendly without adopting predominantly male man-
nerisms like back-slapping and offensive language.

Draw a line on what is reasonably acceptable to you,
such as language and physical contact, and nip inappropriate comments, suggestions, and innuendo in the bud (more
about this later).

Be ready to agree, support, request, and disagree (just
like one of the guys), with a smile that reminds them that
you're a loyal teammate but not one of the guys.

Let your demeanor demonstrate to everyone around you
that you are comfortable with who you are.

Avoid any appearance of impropriety; sadly, there are
always a few men and women in the vicinity who stand
ready to tarnish your reputation.

Remember, your card in the hole (the one that lies face
down where only you can see it) doesn't have to be an ace
to help you win. It just needs to be the right card to fit a
given situation.

Synopsis

Women are accustomed to facing and overcoming obstacles. Recognize the difference between those that arise from
gender-based attitudes, and the ones that apply equally to
men. Then use the key that fits the door you have decided
to open.

Leadership and Credibility
"Which commander is more capable?" – Sun Tzu

Sun Tzu emphasized the relationships between sovereigns and commanders when he said that he would leave the service of a ruler who refused his advice. In the modern corporate setting, every manager or executive needs her superior's support. Otherwise, no matter how skilled or clever she may be, she will lose the challenge. For women, who are more frequently and intensely second-guessed, a lack of upward support impedes their ability to make strategic decisions and their ability to lead.

The Art of War defines several attributes on which the success or failure of a war will be determined, most of which rely on the commander. The most obvious is the leader's capabilities and competence; others, such as discipline, training, rewards and punishment, are the leader's direct responsibility. We might summarize the qualities of a successful leader, and their importance to women, as:

1. Winning the confidence and trust of the sovereign (upper management)

   Everyone needs upward support, women even more so because their authority is more often questioned than that of men.

2. Building and maintaining credibility among the members of your team.

   Women are often better at forming personal relations and reading other people; these qualities can be leveraged into confidence and trust.

3. Ability to prepare, assess, and take appropriate action in a timely and efficient manner (lead).

   To overcome any reluctance to follow their lead, women are well advised to anticipate potential issues before they arise.

Synopsis
To lead with credibility, you will need the support of people to whom you report, as well as those who report to you. This must be earned and substantiated by your performance.

Management and Policy
*"Subjugating the enemy's army without fighting is
the true pinnacle of excellence." – Sun Tzu*

Management gurus Warren Bennis and Peter Drucker define management as doing things the right way. Sun Tzu adds that it is the art of getting things done efficiently — that is, with the least amount of cost and waste. He taught that victory is best achieved through diplomacy and coercion. To this end, the art of deception, self-control and prudence are preferable to aggression.

What does this mean to women?

Is it not obvious how Sun Tzu's philosophy of limiting aggressive behavior plays into women's hands, hearts and character? To emphasize the point, let's take a close look at the Master's advice.

Coercion
Coercion is the use of force, threats, or intimidation to get someone to do what you want them to do (including desisting from things you don't want them to do). It may take the form of threatening punishment in cautioning or disciplining children, of employee rules and regulations (with implicit penalties), of legal requirements, of health risks, and of different forms of conflict.

At first glance, the term seems to convey a negative association, as in forcing, bullying or oppressing. However, it can also serve as a means of pacifying or avoiding still more unpleasant consequences. Admittedly, coercion is a tactic best reserved for intense situations; overuse tends to undermine its effectiveness and to create barriers. Then again, there are degrees of coercion that may range from threats of disfavor and loss of employment all the way up to financial or physical ruin.

It's our experience that, by and large, women favor judicious over heavy-handed methods of controlling their environment. Threats of violence are definitely not the feminine norm – not when women have since time immemorial known that subtlety is more effective and costs far less effort.

Diplomacy

In the time of Sun Tzu, as in today's China and other Asian nations, face is a concept akin to prestige: giving face means enhancing the other person's prestige, or at least not making them look bad. In negotiations, it is customary to word an agreement in a way that allows both sides to gain face.

Diplomatic coercion, then, is where you get your way while creating the appearance that the other side has also won something of value. It is consistent with the win-win philosophy we mentioned earlier.

While coercion can be a useful tool in time of war, managing your work and life extends beyond threats of reward and punishment. Managing is primarily a question of responsibility for administering time and resources to successfully achieve a set or series of defined objectives. It also means coping with difficulties like unreasonable expectations, insufficient resources, and a wide diversity of personalities.

Synopsis

Manage your interests; fight only when necessary and on your terms. Learning how to win without fighting is Sun Tzu's prevailing theme; planning and preparation are the primary tools that make this possible.

Summary

Integrity, reacting to existing conditions, overcoming obstacles, leadership/credibility, and management/policy form the basis of Sun Tzu's principles. Now that you are familiar with these topics, let's move on to the expanded references detailed in the following chapters.

Personal Profile — Khanh Diep
"The momentum of one skilled warrior is overwhelming."
– Sun Tzu

US Army Major Khanh Diep was born in Saigon, Vietnam. Her family immigrated to the US when she was five years old, settling in Houston, Texas. A varsity tennis player, Khanh graduated from the US Military Academy at West Point in 1999, when she married Chad Foster, also a West Point graduate.

Khanh worked as a physical education instructor at West Point, coached the women's marathon team (she clocked three hours 17 minutes in the Boston marathon), and served as a Company Commander in Iraq.

Sun Tzu's theories from *The Art of War* have been very useful to me over the past 10 years. I don't pretend to be an accomplished student of Sun Tzu, but I have had the opportunity to read his classic work, and my husband (also an Army officer) and I have had quite a few conversations on this subject.

If I had to point to one concept from Sun Tzu that has been the most valuable to me, it would be his thoughts on knowing one's self and the enemy. He warns us to "know the enemy, know yourself; [and] your victory will never be endangered." This notion is perhaps a bit idealistic because it is seldom possible to gain a perfect knowledge of one's adversary, and it is also often difficult to make honest assessments of one's own abilities and weaknesses. So many things get in the way such as ego, fear and laziness. However, I have always seen Sun Tzu's idea as having a special applicability to personal problem solving. This is not just a battlefield concept; one can apply it every day.

To this end, I find it useful to substitute "the nature of the problem" for "the enemy." It is vital to look at new challenges carefully in order to fully frame the nature of the problem in one's mind. For example, the problem may involve difficulties in dealing with superiors, co-workers, or subordinates. The better you understand them, the higher the likelihood of figuring out a way of effectively interacting with them. Simultaneously, it helps to have a clear understanding of yourself (strengths, weaknesses, prejudices, fears, etc.). This translates into determining what you can realistically do to solve the issue at hand.

There are many incidents that illustrate this in my life. As a young cadet, I often found myself feeling overwhelmed by the many requirements imposed by the Academy. When I finally took a hard look at what was going on around me and within me, I realized that even the most significant challenges were developmental opportunities, not life-and-death situations.

Daily life as a cadet put my classmates and me into situations where teamwork was essential to success. This was, of course, by design, and I had to look past the stress and chaos of the moment to see the big picture in perspective. This drew me to heart of the issue. I also had to continue making honest self-assessments to uncover realistic options. Sometimes, this meant relying on my friends, or friends having to rely on me. You could say that I applied Sun Tzu's theories both on a personal and collective level with my classmates. Active military duty provided additional applications for strategies from *The Art of War*.

As a company commander deployed in Iraq, I often had to travel through locations identified as hotspots for ambushes and explosive devices. In these areas, the enemy was known to seek "soft" targets, those that appeared less prepared to deal with insurgent attacks. What's a soft target? One where some of the soldiers appear lax in pulling security, even wearing headphones blasting music in their ears! There was at least one report of a gunner on a HMMWV who actually fell asleep during a convoy.

My unit was a personnel support detachment, and we lacked the heavy firepower of an Infantry or Armor unit. Our heaviest weapon system was an M246 Squad Automatic Weapon, a light machine gun with a high rate of fire but not much punch. It was during my preparations for our movements that my thoughts came back to another concept articulated by Sun Tzu, that of making the enemy see "my weaknesses as strengths."

This really got right to the heart of the matter: when my unit was out on the road, we needed to look more like a "hard" target—-one that the enemy didn't want to engage. Fortunately, some of this could be accomplished through the maintenance of solid military discipline: soldiers vigilant and sharp at their stations with weapons at the ready.

One tack I initiated was to tuck our convoys in with larger, better-equipped elements, such as Military Police or Combat Engineer units, whenever possible. When we had to move on our own, I beefed up our firepower by obtaining a few heavier (50 caliber and M240B machine guns) and made sure that my soldiers were properly trained on them. In other words, where I assessed my unit to be vulnerable, I took action to strengthen our skills and armaments to increase the appearance (and reality) of strength. This sent a clear message to the enemy that we weren't going to be an easy target.

Perhaps the greatest lesson I learned from *The Art of War* was to gain control of whatever situation in which I found myself. Sun Tzu doesn't imply that you can be in complete control, but he does suggest that it is possible to influence conditions to your favor. In today's military lexicon, this is known as "shaping the battlefield." And once again, the strategy extends well beyond combat into everyday life.

As mentioned earlier, it is vital to understand yourself, especially your weaknesses. From there, you need to work toward turning these weaknesses into strengths. This concept isn't exclusive to Sun Tzu, but it has inspired me to upgrade the areas where I am less comfortable. Skill is improved not only by focusing on what you are already good at; instead, you need to reach outside your comfort zone to work on areas of weakness. This not only makes you better individually, it gains respect from peers and subordinates.

Let me tell you, earning respect is especially important for a woman in the military. Even though the US Army is probably the most diverse and accepting institution in our country, it is still heavily dominated by men. Above all, however, it is a meritocracy. Working through intimidation to earn your place in a unit isn't exclusive to females. But I believe that it is magnified for women due the traditionally subordinate role we've played both in the military and society in general. When your colleagues see that you are willing to face your fears and work to improve in areas

where you are clearly uncomfortable, you will earn
the respect that is essential in a military or civilian
environment.

Serious students of Sun Tzu may focus on his more
nuanced views concerning strategy and combat, but
we can't forget that these thoughts have applicability in many other areas of personal endeavor. Not
every challenge involves dealing with bullets, but just
about all of them require one to overcome stress and
to resolve problems either individually or as a team.
I can think of few better sources of inspiration than
Sun Tzu.

2. Preparation

*"No country has ever profited from
protracted warfare." – Sun Tzu*

Sun Tzu's second chapter explains his strategy for employing military forces. Here he emphasizes efficiency, timing,
expenditures and motivation. These issues have many close
parallels with women's concerns in the competitive world
of business.

We've all heard corporate and athletic figures refer to
beating their competition as "going to war" in order to
evoke a spirit of aggressive effort. A healthy alternative
strategy is to emphasize a more positive and practical tone,
that of winning with the least amount of loss to either side.

Military, business and private operations all require adequate budgets to provide the human and other resources
needed to prevail. This can include full-time employees,
consultants, equipment, transportation, provisions, potential backup, and many other needs. The larger the effort, the
more costly; and the longer it lasts, the more difficult it will
be to sustain in terms of cost, effort and morale.

What does this mean to you?

We recognize that women often struggle to compete on equal terms with men within the corporate world. And when given the opportunity to do so, a brighter spotlight is likely to be focused upon them. This results in tighter scrutiny, less tolerance for error, and harsher criticism than might be brought to bear upon males in similar roles of responsibility. And so, in order to get credit for doing a good job, a woman needs to be well organized. This begins with effective planning and resource allocation.

Assess Your Resources
"...a speedy victory is the main objective." – *Sun Tzu*

Assess
Before committing to a major effort, make the following assessments:
1. Can it be accomplished within a reasonable amount of time?
2. Will you have sufficient resources available for the time and effort required?
3. What contingencies will be available if it costs more/ takes longer than expected?

Do not accept responsibility for a plan that is:
1. under-funded;
2. understaffed;
3. scheduled to be completed more quickly than you consider reasonable; or
4. in danger of being extended for an unusually long period.

Don't let anyone try to pressure or intimidate you into agreeing to a poor plan, inadequate resources, and/or unrealistic milestones. Otherwise, the project is likely to fail, and you'll be left holding the proverbial bag.

What To Do

If you are getting resistance from your boss about the amount of time and resources needed to accomplish a milestone or objective, break the project down into fine detail to make your point. You may find Microsoft Project or another planning software tool helpful to this end, and it will also enable you to assign tasks and follow progress once you get started.

If the undertaking is likely to last upwards of eight or ten months, try to separate it into two or more separate deliverables that may involve some different staff members. There are several reasons for this:

Over time, delays tend to accrue and frustrations intensify, leading to exasperation, sloppy work and fading interest. Some personnel changes are to be expected between different projects but are inconvenient during the course of a deliverable.

Some of the inevitable mistakes you make early on may be avoided thereafter.

Completing a defined task builds confidence and is personally satisfying; it also scores more points than just finishing one of a series of deliverables.

Regarding gender, it's usually better to have a ratio of males and females on your team that is representative of your company, rather than loading up with women or avoiding them altogether. Because, fair or unfair, people tend to notice things like this when a woman is in charge.

Synopsis

Determine for yourself if you have sufficient resources to get the job done; don't take anybody else's word for it.

Plan Efficiently

"To be prepared for war is one of the most effectual means of preserving peace." – George Washington

Preparation doesn't guarantee success, although a lack of preparation comes close to assuring failure. Men have a tendency to forge ahead with more energy and optimism than foresight and caution. Solid preparation can offer women a competitive advantage by avoiding predictable errors, delays, cost overruns, and a whole bucket of potential pitfalls.

Like potholes, small problems become big ones when left to themselves: loose bolts lead to engine failure, rot fells massive trees, and mere suspicion has been known to wreck careers.

RR was a fashion designer for a prominent dress producer in New York City. Her designs sold well when properly cut and assembled, but sold poorly when the workmanship was shoddy. Since she was ultimately responsible for the success of her line, she became involved in the details of the seams, buttons, zippers, pricing, and many other aspects. As a direct result, her dresses captured a significant portion of the marketplace.

From this RR concluded that a good design is worthless without attention to the details of how the garment is produced and sold.

Efficient planning usually includes considerations like a clearly defined product or service, scheduled deliverables, adequate resources (staff and funding), upper management support, good communications, and contingencies. Bear in mind that company budgets are subject to change, people come and go, and whatever is hot today may eventually fizzle. As mentioned earlier, don't fall into the trap of accepting unreasonable schedules and inadequate support.

Synopsis
Murphy's law cautions that "Whatever can go wrong will go wrong."

To which we amend, "But only if you let it."

Everything is subject to change; if left alone, conditions tend to deteriorate. In order to exercise a degree of control and guidance over significant changes, you need to be alert to the signs before, during, and immediately afterward.

Build on Strength

We all start small.

Separating large projects into smaller deliverables is like setting bricks as a foundation and then building upward. Your strengths include your demonstrated skills plus your validated accomplishments. These grow in scope as you develop more experience.

If you're technical, develop your administrative and overall people skills; otherwise, seek help where you are most vulnerable. You don't need to be an expert at everything to effectively manage projects and lead people, but you do need to be realistic about your strengths and weaknesses.

Nor should you try to hide your limitations, because they will become obvious to your team members if they aren't already. Males may attempt to cover or disguise their deficits with bluster, whereas women are more inclined to internalize and agonize over their perceived shortcomings. A better tack is to let certain individuals know you are relying on them, and to acknowledge their contributions directly and to others.

The best way to strengthen your weaknesses is to view them as self-improvement projects. Remember our discussions about yin and yang? Clearly, strength and weakness provide another example of this balance. Since women often find it easier to accept their limitations, why not view them as degrees of strength — in other words, as qualities to be upgraded?

For example, let's say you're stronger in communication skills than you are with a computer. What should you do?

If you need a working knowledge of commonly used software, take a course to get started in the right direction (many classes are offered online and can be worked on at the student's convenience). Chances are you won't need to become an expert, only to develop a working knowledge.

Which lesser strengths do you need to fortify? The ones that are holding you back where you are employed, or those that are keeping you from getting the kind of job you really want? If you are in doubt about this, ask a colleague and your boss. One approach might be:

"I would like to grow within this company and would appreciate your advice as to which skills I might improve in order to make a valued contribution."

Make a list of these lesser strengths and rate them by:

1. importance to your career;
2. level of difficulty and time/effort needed to strengthen them.

Obviously, a desirable but non-essential skill which is likely to take a year or more to master will be a lower priority than a tool that is critical to your work right now.

Synopsis

You are the ground on which you stand.

Provide Value

...and get credit

Providing value is straightforward enough to require little explanation: you do what is asked of you, do it well and in a timely fashion, and find other things to do that are also needed. Since you are being paid to provide value through the results of your assigned tasks, it seems reasonable to make the effort.

Perception

However, being recognized for the value you provide isn't as simple as it should be. People tend to observe what they expect to see. What do they perceive when they look at you?

To succeed at work, it isn't always enough to do a good job and wait to be recognized: you must be perceived as providing value. Not to belabor male-female stereotypes, but men seem to be more adept at advertising their abilities and accomplishments than women, which means that they may get more credit for a lesser job. What can you do about this without climbing atop your desk and tooting your horn like a political candidate?

If you happen to have a boss who recognizes your contributions and lets others know about them as well, lucky you. But this doesn't mean you can afford to disappear into your cubicle and leave your career up to him/her. Think about it: your current supervisor may leave or assume another role, and the next person to whom you report to might see things (and you) differently. In the real world, it doesn't pay to leave your career in the hands of anyone who isn't you.

What exactly is perception? Viewed through the window of yin and yang, it is the balance between image and reality. Most of us realize that when a magician saws someone in half, she or he is deceiving our perceptions. We don't really believe that flesh and bone are being severed then restored to their natural state. Hopefully, we understand that a sleight of hand is being performed for our entertainment.

However, many men and women are taken in by con-people who prey on the most innocent and naïve marks they can find. Millions have invested in non-existent stocks, provided access to their credit cards and bank accounts to criminals, and trusted people who are out to swindle them. They are seduced by false impressions.

Control

In the workplace, false impressions may be difficult to recognize and deal with. To exercise at least some control over the impressions your colleagues and managers have of you, you need to be aware that:

1. in any work environment, everybody forms an impression of everyone else;
2. impressions are usually based more on hearsay than personal evaluations, especially concerning women;
3. some individuals actively spread rumors to influence impressions;
4. passive people exercise less influence over the impressions others have of them.

In other words, the value of your contributions may not be universally appreciated or recognized. And for female employees, this can be a more delicate issue than for their male counterparts because women aren't always valued the same way as men.

In reality, both male and female employees encounter their share of unfairness. Men harbor lower expectations of being treated fairly and realize that they need to compensate. Many women begin by expecting justice and, when it does not occur, interpret this as the result of personal weakness or gender bias, which may or may not be the case.

Our point is that in the workplace, as elsewhere, fairness cannot be taken for granted. There are people who will happily take credit for your accomplishments and push you aside without a speck of guilt. Whether motivated by personal gain, desire for power or a malicious streak, they play their game when and where they think they can get away with it. Like bullies, they target victims whom they perceive as weak and reluctant to stand up for themselves. Some – males and females alike – may believe that women are easier targets because they tend to avoid confrontation.

If it suits your style, by all means address someone you have reason to think is spreading negative impressions about you. One approach might be to ask (in private),

"I was wondering if there is any communication problem between us. I'm sure that if we get to know each other better, we'll be able to clear up any misunderstandings." or

"I've never had the chance to get to know you – are you free for lunch? "

These approaches are likely to change their attitude toward you. Even if your approach backfires, they'll know that you're not going to be a passive victim.

Of course, direct confrontation is not the only way of dealing with bullies and other unreasonable types. For instance, a little networking can be quite effective, like becoming friendly with an influential man or woman you respect (more on networking later in the book).

Example:
"When you have a few minutes, I'd like to share with you my thoughts about XYZ, and what I might contribute to this effort." (You could then support your ideas with examples of your accomplishments.)

If your boss or another respected individual speaks well of you, the value you provide is far more likely to be recognized.

Another tactic for controlling people's perceptions of you is in the way you talk about your colleagues and associates. Few people seem to recognize the fine line between discussing colleagues and spreading gossip. Bottom line: everything you say about someone in the workplace will get back to them (negatives twice as fast). In the final analysis, what you say about another person reflects more on you than them. So praise when possible, and be no less supportive behind their back as when they're present. If asked about a person you don't like or trust, try something like,

"I've heard she is a strong administrator" or, "I haven't had the opportunity of working very closely with him."

Most people will respect you for not saying anything negative.

Synopsis
The value you provide is real; perceptions are subjective. Take a proactive role to assure that you are recognized for your accomplishments.

Exercise Control
Excessive power = loss of control.

Most of Sun Tzu's references to power are in relationship to authority (being able to issue orders), deception (pretending to be more or less powerful to confuse the enemy) and timing (exercising power efficiently without waste). Control is promoted throughout *The Art of War* to position forces and resources, and to avoid following (the ruler's) instructions when necessary.

Power enables you to make decisions and act upon them; control can be used as a mitigating influence to weaken this ability. In other words, power and control are opposite yet complimentary forces, much like yin and yang.

In functional terms, power is directed primarily toward the result, whereas control is focused rather on the course of action used to achieve the result. Either can be used by you or against you.

Wielding Power
The most judicious pursuit of power is to recognize your available options. However, just as Sun Tzu teaches us that the greatest victories are achieved without fighting, the wisest use of power is not to exercise it.

If you possess great power, it would be heavy-handed (and unnecessary) to throw your weight around like a professional athlete knocking over amateurs. On the other hand, small amounts of power may be precarious to put to the test, where they can be challenged or denied.

Either way, abusing power raises more resentment than cooperation, and it also motivates others to create controls against you.

The clever way to apply power is with subtlety, without anyone realizing that you're doing it. Sound familiar? Sure, since women have been successfully applying this tactic since the beginning of recorded history, and probably a whole lot longer! How best to do this? Let us count the ways.

A few examples:
1. Convince: The power of persuasion is enhanced by your standing (position) in the company or other environments. People are generally more willing to be convinced than ordered, especially by someone whose power they cannot refuse.
2. Request: Asking someone to do something is a polite alternative to telling them to do it. Those who recognize your power most vividly will appreciate your lighter touch the most.
3. Bargain: If you can offer a favor or privilege, even if it amounts to little more than gratitude, you are likely to gain respect and loyalty.

Restraining Power
The primary role of control is to restrain power.

Organizations try to limit power with required procedures, accountability, security, and of course consensus. Committee meetings and multiple approvals have probably stopped more projects from going forward than prison walls have kept criminals locked inside.

In fact, the most intense and ongoing battles fought within the business world are between power and control: who gets to decide and do what? Without controls, the power seekers would eventually run the world, adding chaos to the damage they already do. The controls they try to impose are intended to consolidate their power; the controls the rest of us need to apply are to limit their power and, by extension, their control over our lives.

We need look no further than the dictators of the past century as examples of power addicts out of control. Yes, these are extreme cases to be sure, but even they started on a small scale. The old saying "Power corrupts, and absolute power corrupts absolutely" refers to power uncontrolled.

We're not suggesting that you, single-handedly, are responsible for restraining those who throw their weight around, but you do need to protect yourself from falling under their control. Whether at work, home, or somewhere in between, be conscious of this balance in your own behavior, and in those whose concept of the world is not the one you want to live in.

Synopsis
Exercise what control you can to mitigate the power others have over you; and balance the power you may have over others with a thoughtful dose of self-control.

Summary
Preparation — homework, if you prefer — is the difference between maximizing opportunities and wasting time. In order to apply what you've got going for you, you need to accurately and realistically assess your resources. Remember, the time to check whether your parachute is in proper working order is before you jump. Nor can you leave this vital issue to someone else's care, or to faith.

Preparation lays the groundwork for efficient planning, proper timing, and controlling costs, schedules and resources. This, in turn, will position you to anticipate and resolve problems, take advantage of opportunities, and excel as a manager and a leader. In today's business environment, a woman may be expected to outclass the competition in order to be afforded equal terms.

3. Strategizing

"If you know the enemy and yourself, you'll survive a hundred battles." – Sun Tzu

Try immersing your left hand in a bowl of hot water and your right in icy cold water, and then dip them both into a larger bowl of water at room temperature. Now that both hands are exposed to exactly the same temperature, a strange thing happens: the left hand feels cool, and the right one feels warm.

Similarly, a 70-degree afternoon is likely to feel cooler to a Sahara desert dweller than to those who live in colder climates. So they would be wise to pack and plan differently before traveling to the same destination.

We could go on with examples of how different societies, individuals, and even body parts subjected to the same conditions can experience completely different sensations. This is why an office environment presents different challenges to the people who work there, and why an understanding of yourself, your competition, and your clients is essential to your success.

How does this work for women?

Recognize Your Assets

In the realms of bookkeeping and accounting, assets are the economic resources owned by companies, institutions, families and individuals. The two main classes of assets are known as tangible and intangible.

Tangible assets include available cash and valuables (like property and inventory) that can readily be converted into cash.

Intangible assets consist of non-material resources like trademarks, patents, copyrights, accounts receivables, and reputation. For individuals, they extend to personal qualities (like being a good listener), accomplishments, experience, recommendations and such. Women are more often aware of and sensitive to these abilities than men, which can enable you to use them to advantage.

For our purposes, we'll sub-divide the intangibles into accomplishments (degrees, certifications, track record et al) and personal qualities, like being:
1. a good listener and communicator
2. patient and energetic
3. supportive
4. a reliable team player
5. respected
6. positive and upbeat

Combining several of these qualities can position you inside the comfort zone of colleagues and managers. Given at least adequate tangible assets, this can make people more willing to share confidences and opportunities.

When you are entrusted with responsibility for projects or deliverables, your assets are the resources available to you. Recognition of these assets is the first step toward understanding yourself.

Synopsis
The strongest nails are only helpful if you know how to use a hammer. Value your strengths and shore up your lesser strengths.

Overcome Your Limitations

Assets and limitations are the yin and yang of your ability to get things done. In order to capitalize on your assets and minimize the downside of any limitations, you need to recognize and acknowledge them.

Many women have a tendency to underestimate their abilities and agonize over what they believe to be their weaknesses. A better way of dealing with the latter is to view them as lesser strengths subject to improvement. Then strive to make improvement a reality.

Example:

If certain technical skills are needed to contribute to your project or deliverable, bone up in the area of concern and enlist assistance from someone who can help you. One possibility is to request the required expertise (no one is expected to master everything); another might be to offer something of value in exchange for their help.

Traditionally, male egos tend toward outward shows of confidence, whether or not that confidence is real or warranted. Women, on the other hand, are all too often self-effacing. False confidence and bluster vs. self-effacing modesty: extremes of yin and yang. Neither extreme is likely to inspire trust among your colleagues or managers. Quiet confidence, born of self-awareness, stands strongest and endures; likewise, recognition of a particular limitation you are in the process of addressing is more likely to win respect than scorn. Balance lies somewhere in between, more an ebb and flow than a specific place.

So the best way to overcome your limitations is to:

1. acknowledge them (at least to yourself)
2. create a plan for improving them
3. compensate, when necessary, by leaning on your strengths and seeking help with your weaknesses

Don't get down on yourself for not being perfect, because no one is.

Synopsis
Perfection isn't necessary as long as you are competent. This means knowing how to achieve objectives that may require skills you lack.

Analyze the Competition

*"If you know yourself but not the enemy, you will
sometimes win, sometimes lose." – Sun Tzu*

Too much focus on your perceived faults can blind you to the people with whom you may be working and competing. Remember that you aren't functioning in a vacuum, but among other imperfect men and women.

With an image of your personal assets and liabilities in your mirror, it's time to analyze the competition. This may be someone within your group who is vying for the same promotion, or people in other companies who are targeting the same clients. This process may give women an advantage over their male counterparts.

Why is this so? Because women are generally known to be better listeners. Not all, not always, but more often than not.

The common observation that men don't listen to—or hear—what women say is often true. However, it should also be mentioned that not all women are great listeners, and some men really are.

The title of Allan and Barbara Pease's *Why Men Don't Listen and Women Can't Read Maps: How We're Different and What to Do About It* feeds several stereotypes that are both instructive and potentially misleading. Rather than debate the accuracy of these claims, let's see how we can use them to advantage.

The happy fact is that the overwhelming majority of women can learn to read maps quite well, even those who may have thought they couldn't. When you grow up being told that you're better at some things than others, it's natural to go with the flow. Fortunately, it doesn't take an overly rebellious nature to oppose these gender stereotypes, only an open and curious mind.

To be fair, many men can learn to become better listeners. The trick is to let them know when their opinion is wanted, or when you just want them to listen. Some guys actually get it, sometimes[*].

Why all this talk about stereotypes? Because your competition probably consists of both men and women, and an understanding of their tendencies is essential to your analysis. In other words, be mindful of common gender-based stereotypes without being blinded by them. Consider as well that the same stereotypes can influence the way people look at you.

Place the focus of your analysis of the competition on their effectiveness: are they successful, and how do they go about their business? Are they hard working and well prepared; charming; smooth talkers? Do they offer substance or smokescreens? Do they honor their commitments or resort to excuses? Analyze their assets and limitations, and use the information to improve your ability to compete with them successfully.

Examples:
Competitor #1 is experienced and very smooth, but your product or service is superior. In this case, focus your attack on differences between the products, not on personalities.

---

[*] Try giving him a copy of John Gray's 1992 publication "Men are from Mars, Women are from Venus."

Competitor #2 has less than exemplary communications skills. So you create a concise, articulate synopsis of your status and have it ready for the boss each Monday morning.

Competitor #3 seems to hang around the boss's office looking for opportunities to ingratiate herself. You might make a point of showing up early once a week to offer a verbal status synopsis, and to ask the boss if there is something important with which you could lend an extra hand.

Competitor #4 (a male) is clever with acronyms and numbers, which are impressive at the weekly status meetings. Your strategy is to bone up on all the facts and figures relevant to your work, and to present them clearly and confidently at these meetings. Given common stereotypes, chances are, you (a female) can get higher scores for equal performance.

Remember, the more you know and understand about the competition, the better you can position yourself to come out on top.

Synopsis
Observe the competition carefully, and seek advantages. The best thing about competing with others is that you don't have to compete with yourself.

Set the Odds in Your Favor
The overwhelming majority of people are adverse to risk, and yet they take risks almost every day. The reason for this apparent contradiction lies in their inability to accurately assess the risks they take.

In battle, larger numbers and superior materials can give one side an obvious advantage over the other. In more peaceful environments, seek a competitive edge through strength of knowledge, preparation and execution. Playing against the odds will fail more often than not.

We've all heard the suggestion "Let's give it a shot and see what happens." That's kind of like basing a decision on a coin-flip, which may be OK if you're considering the cheaper stapler or roll of tape, but you probably wouldn't use this approach to buy a car or make an important job or career decision. In order to improve your chances of success, the risks you take on need to be clearly calculated.

What is a calculated risk? One where you have thoroughly considered these key factors:

1. Probability of success.
   The odds against winning most state lotteries are millions to one. Yet people buy tickets by the billions. Their reasoning, of course, is that spending a few dollars for the possibility of becoming rich is an affordable risk. In the workplace, spending a few hours preparing a proposal may be worth the effort even if the odds do not favor its acceptance. Depending on the nature of your ideas, the potential downside may be wearing out the welcome mat and losing credibility.
2. Potential risk vs. gain: does the benefit justify the risk?
   Experienced gamblers will tell you that putting up a large sum of money in order to win a relatively small amount is a poor bet. Yet office workers constantly risk their credibility on ventures which, even if successful, stand to gain hardly any benefit at all.
3. Factors likely to influence outcome.
   The key to assessing essential factors that can make or break a deal for you are whether they are known to you, and the extent to which you are able to exercise control over them.

If you suspect that there may be things going on behind the scenes, you lack the basis of evaluating the odds of success or failure. Likewise, if the outcome will depend in large part on decisions made by others, the project may be out of your control no matter how hard and diligently you work on it.

In either case, you need accurate information, as well as commitment and backing from someone in a position to strongly influence the venture. Never assume that others are operating in good faith or judgment.

Given a satisfactory answer to these issues, the next step is to create a detailed evaluation of:

1. The desired result: To begin with, is it doable, or more likely a pipe dream? If successfully developed and implemented, is it likely to function as expected? What do your colleagues and managers think about it?
2. Resources needed: Staff and funding are among the key ingredients in any project. You need enough skilled individuals to do the work, and adequate funding to allow for proper facilities and time constraints.
3. Schedule: A major parameter of success in any effort is whether sufficient time has been built into the deliverables that define the project. After all, not even Peoria was built in a single day. Unreasonable schedules, which strongly influence funding, are responsible for the majority of project failures.
4. Your role: Will you be assigned to lead the project? What precisely are the factors on which your performance will be judged? To what extent are these factors under your control?

These considerations are based upon evaluating major company projects. However, they also apply to personal decisions, such as whether to get involved with certain projects, maintaining credibility, networking, forming close professional relationships with colleagues, and the like.

Synopsis
Instead of looking for a way to beat the odds, make them work for you.

When you gamble, expect to lose.

Get Ready to Win
"If you know neither the enemy nor yourself,
you'll be defeated every time." – Sun Tzu

Preparation
Lucky people are the ones who do their homework. This oft-mentioned theme, derived from the writings of Sun Tzu and countless other teachers, is a key step in the pursuit of success.

Here again may be found subtle advantages for women. For while males are sometimes tempted to shoot from the hip and fly by the seat of their pants, women tend toward a more patient and cautious approach. This is consistent with thorough preparation.

If you're early in your career or in an unfamiliar environment, one of the best ways to prepare for an endeavor is to record as much of it as possible in a planning software format like Microsoft Project. Such tools are useful not only for managing complex projects, but also for predicting and tracking the responsibilities and productivity of a single group or individual.

Another form of homework is to familiarize yourself with tools, programs, systems, and procedures relevant to the work you will be doing. If you're not a programmer, you probably won't be expected to learn the latest codes and protocols, but it wouldn't be a bad idea to recognize internal acronyms and have an overview of how your company's systems interact. Regard your strengths as functional assets, and your weaknesses as lesser assets to be strengthened.

Preparation, then, is essential to deciding whether to get involved with a project (given the option), and also getting ready to contribute to it effectively.

Commitment

Once you have agreed to participate in a project, team effort or individual responsibility, set your mind to doing it as well as possible. Half-hearted efforts are doomed in process and effect: people around you notice and lose respect for you, and your results will obviously suffer. Better to resign and seek another job than spin inevitably downward into oblivion.

Anytime you're feeling overwhelmed or are having trouble getting something done on time, ask your supervisor for assistance. Tell her/him that you are committed to your role, but need some help in getting up to speed or catching up. Understand that men run into problems and blank walls no less often than women – it's the way you face and overcome a problem that sets you apart. The point is that as difficult as it can be to admit to such problems, doing so is better than falling behind or making serious mistakes. In fact, most managers prefer this type of honesty up front to having problems stack up on their doorstep later on.

Visibility

In warfare, Sun Tzu recommends deception in dealing with one's enemies. In the workplace, however, deception should be applied with a light touch. After all, no matter how intense the competition, you're probably not facing life-or-death conditions, and you don't want to develop a reputation as a liar or phony among allies or opponents. This is not to say that you are obliged to reveal all your plans and preparations to everyone around you. As Sun Tzu said, "A leader...varies his techniques and strategies to keep other people from recognizing them. He shifts positions and takes indirect routes to keep other people from anticipating him."

In the modern workplace, a woman should consider how to apply the Master's advice internally (to colleagues) and externally (to outsiders).

In your place of work, openness and consistency are likely to be encouraged over secrecy and unpredictability. Still, any negative feelings or anxieties you may feel toward fellow workers, as well as certain job-related or personal situations, are best kept to yourself. Surely you have noticed how many men equate revealing anxieties to a sign of weakness.

Bear in mind that a secret shared is no longer secret. Among your colleagues, tell them only what you want them to know: keep your doubts to yourself, and choose your confidents carefully. With regard to those who can hurt you, try not to give them reasons to single you out as their enemy, and watch them closely.

When competing with other companies and their representatives, you might prefer to keep them off-balance and guessing about your new products, services and policies. You most certainly would not provide them with a list of your existing or potential clients, or any special promotions and discounts in the offing. Indeed, the less they know about your plans, the better.

Efficiency

One of Sun Tzu's most powerful and persistent messages to us across the ages is to win efficiently. That is to say, quickly and with a minimum of expense and loss. His basic principles are to:

1. know where, when and how to apply your resources;
2. make sure that all the participants in an endeavor are on the same page;
3. make an effort to prepare for the task, and
4. be sure that your responsibilities are clearly defined and understood.

Synopsis
Winning is primarily about preparation, attitude and tactics.

If you don't know what you're doing, you'll never get it done.

Summary
Every office environment poses a distinctive, yet familiar set of challenges. And every individual brings her own unique complement of skills and experience to the table. Sun Tzu tells us how to build these factors, and along with an understanding of the competition, to create a successful working strategy. This means going with, not against, the odds in order to achieve your short-and long-term career objectives.

Personal Profile
Tammy Rosenthal — Stay focused on your goal.
Dr. Tammy Rosenthal, a graduate of George Washington University Medical School, practices medicine at Kaiser Permanente Health Care in Reston, Virginia. She grew up in New York's Westchester County with a French mother and an American father.

Sure, I have worked pretty hard to get where I am today, although I'm not so sure I think like a warrior. In fact, I've probably missed a bunch of opportunities that more aggressive and ambitious people might have grabbed. Nevertheless, I've learned a thing or two from Sun Tzu's lessons in *The Art of War* and have used them to good advantage throughout my life and career.

Growing up, I was (and still am) very lucky to have awesome parents. They supplied the tremendous advantage of resources to send me to great schools and expose me to the arts, music, sports and travel. Now I have a great job (except for 2 AM emergencies), earn enough to take care of my family, and have time for other interests. Mom and dad also instilled in me the expectation of getting a higher education, pursuing a successful career, staying out of trouble, and giving back to the community.

Thinking back, it seems natural to associate Sun Tzu's influence with that of my parents:

Sun Tzu stressed the value of staying focused on your goal. Mom would not have had it any other way.

Variation in tactics: My younger brother and I were sent to France for a summer at the tender ages of seven and ten with a very limited knowledge of the language; we both learned to think on our feet and to learn on the run.

Flexibility in your methodology: Don't laugh, but after our TV broke, dad realized my brother and I were reading more, so he definitely took his time to fix it.

Formulate your plan of attack: In my case, that amounted to twelve years of dedicated higher education with the encouragement and support of my family.

Analyze the enemy: In my profession, the enemy is whatever illness or condition I may have to treat. Some problems require direct attack, while others respond better to more subtle treatment.

Focus on your goal: Before becoming a doctor, my goal was to become one; as a doctor, my objectives have broadened toward personal improvement and providing the very best possible patient care.

Sun Tzu stressed winning without fighting; I recognize this as a direct parallel to preventative medicine (teaching and helping patients to avoid injury and illness).

Lead with confidence: Treating patients requires both an inner confidence (that you're doing the right thing) and the ability to gain their trust.

4. Allocating Resources

*"Invincibility lies in the defense; the likelihood
of victory in the attack." – Sun Tzu*

Sun Tzu equated invincibility (being unconquerable) with
self-defense, and the potential for victory with recognizing the opposition's vulnerability and knowing where and
when to attack. Preparation, identifying vulnerabilities and
timing – these tactics are familiar to successful women.

What's the message to a woman? Quite simply that you
learn to marshal your resources and use them to maximum
advantage. This time-tested strategy is based on observation, objectivity and patience.

Believe in Yourself

If you don't, who will?

Believing in yourself begins with an assessment and acceptance of your strengths and lesser strengths. Generaliza-
tions about the differences between males and females suggest that men may be inclined toward optimism about their
strengths while downplaying their weaknesses, whereas
women all too often lean in the opposite direction. This why
you are advised to be objective in appraising your personal
assets in order to apply them where they can work for you,
and not be paralyzed by your perceived limitations (assets
under development).

Those of us who aren't perfect are subject to areas of
weakness relative to any given situation. If you need to deal
with Chinese business people and are unfamiliar with the
language, you can either learn Chinese or use a translator/
interpreter. Is it worth your time and effort to attempt to
master Mandarin? That depends upon the extent of expo-
sure you anticipate, and your motivation to take on a very
ambitious task.

The same logic can be applied to programming codes, biochemical components, engineering terminology and telecommunications networks, to offer but a few examples. Generally, an overview of the applications and terminology that may affect or overlap with your work is useful in establishing credibility and understanding. Some of the most brilliant specialists in the world are poor managers, and solid management skills do not necessarily require heavy technical abilities. The most valuable common denominator is the ability to communicate across the scope of your responsibilities.

If you feel that you need to bolster your self-confidence, here are a few pointers:

1. Base decisions on your personal experiences; the ones that worked least well are no less valuable than your successes.
2. Establish short and medium-term goals to provide a sense of direction.
3. Analyze your least successful results to understand what went wrong.
4. Reaching for the moon is OK as long as you don't expect to get there; establish a balance between ambition and what is realistic.
5. Listen to criticism as if it were directed toward another colleague. If it makes sense, learn from it; otherwise, consider the source.
6. Provide support and constructive feedback to your colleagues and associates; some of them are likely to respond in kind.
7. Give your time and energy to others. When you do this, you will get positive feedback and respect from others. These are building blocks for self-respect, which is essential to believing in yourself.
8. Never get down on yourself! Errors are the seeds of

learning and eventual success, as long as you do not repeat them.

Synopsis
Lead with your strengths, improve your weaknesses, and move forward.

Visualize Success
Win before the battle.

Sun Tzu's philosophy of winning implies visualizing details of the forthcoming encounter from preparation to engagement and ultimate victory. In this way you can create a positive image of winning prior to the effort. We are, of course, referring to preparing, planning, and carrying out your agenda, not sitting back and fantasizing fame and glory.

By and large, women are more practical than their male counterparts. For this reason, women are more inclined than men to wait and see what happens. To the maxim that doing is better than dreaming, no argument. However, visualizing—that is, planning out the steps and stages of a game plan—can reveal needs and problems before they pop up to bite your nose.

Visualization is not the exclusive realm of artists, philosophers and lottery hopefuls, but as well a contribution to any important undertaking. Albert Einstein referred to imagination as "...the preview of life's coming attractions." Obviously, not everyone is capable of Einstein's vivid visualizations, but a little imagination can provide inspiration and enthusiasm.

Creating a positive (as opposed to tentative or negative) aura within yourself and among your colleagues is an often overlooked advantage. Never underestimate the power of an upbeat attitude, especially where hard work and a sustained effort are called for.

If you're having trouble getting started, try the following:
1. Focus on a single goal, like learning how an important computer system at your company functions.
2. Draw a mental picture of yourself smiling as you accomplish this.
3. Imagine the feeling of success alongside the image. Aren't you proud of yourself? Doesn't it feel good enough to justify the effort?
4. Repeat this process every time your enthusiasm, or imagination, begins to fade.
5. Be on the lookout for negative visualizations, i.e., the kind that focus upon difficulties and potential (or past) failures. Then replace them with upbeat versions.

Synopsis
Form an image of where you want to be, and extend the borders of your goal as you approach it. As you grow, so must your agenda.

Reduce Mistakes
"Intelligence is the ability to learn from your mistakes. Wisdom is the ability to learn from other people's mistakes." - Ruhtra

During childhood our brains are wired to memorize and associate information. As we mature, we gain the ability to acquire and apply our knowledge to the tasks of living. The application of theory (what we've learned) to accomplish simple and complex goals creates experience.

Sun Tzu understood that mistakes are spawned by faulty judgment based on incomplete or unsound information, lack of preparation, insufficient experience or skill, poor leadership, laziness and indifference. The more elaborate and ambitious the undertaking, the greater the risk of making a mistake; the more mistakes, the greater the window for failure.

It has been said that only sure way to avoid mistakes is

to do nothing. If, however, you are paralyzed by the fear of making a mistake, you have already lost the battle. Success requires action and a degree of risk-taking; mistakes play an essential role within the learning process. The imperative is to avoid major mistakes – the ones that cause you grief – by visualizing and properly preparing your course of action.

One practical step is to learn from past mistakes and avoid repeating them. This requires an honest, maybe even gut-wrenching review of what went wrong, why it happened, what you might have done to avoid it and, most importantly, how you can keep it from happening again.

"I was petrified by fear of making a mistake," architect Gail G. tells us, "until I realized that my male colleagues made mistakes all the time. For the most part, they just brushed them off and pushed ahead.

"At first I had to force myself to take even small risks, even though I realized that otherwise I'd fail for sure. With practice it got easier, and now I'm known as a 'gutsy gal' among my predominantly male colleagues. Really, I'm no more confident than they are, but they look at me differently because I am a woman. So I just smile and reap the glory."

When something does go wrong, be sure to acknowledge responsibility for your role in whatever happened; conversely, do not accept blame for someone else's blunders. Once you've come to terms with what you did wrong (or failed to do right) you will have learned at least two important lessons: that you are an honest and responsible person, and how to avoid repeating the same error.

Learning from the experience of others is like watching someone step on a land mine, alienate a colleague, or allow their health to deteriorate. These aren't things you need to do yourself to figure out why they are bad ideas.

In the end, the side that makes the fewest mistakes will increase their chances of winning and avoiding criticism. This certainly applies to women.

Synopsis
Even the most carefully calculated risks engender errors.

Do not fear to make mistakes; learn from those you make and the ones you see others make. This is the basis of experience.

Act from Strength

Leverage is borrowed strength.

We've all heard about the 500-pound gorilla that sits anywhere he wants. None of us is likely to outmuscle that guy, so we must find another way to deal with him. And since we are presumably smarter, our most likely course is to outthink him.

In the conventional work environment, individual strengths and weaknesses are more subtle. Unless you work with someone closely, you may only have a vague idea of his or her strong points. That's why you need to rely on your own strengths.

These may be personal qualities and abilities, or those available to you through colleagues and people on your team. If you are competing with another group, individual or product, the nature of your assets is relative to the competition's strengths and weaknesses. Learn to leverage your strong points – whether listening skills, superior preparation or whatever.

Examples:
1. The boss lets you know that she received an impressive written proposal from your major competitor for a key assignment. So you create a PowerPoint presentation to display your artistic and convincing verbal skills.

   Once you have her attention, you ask how you might have made your presentation more relevant and effective. Then you include the changes she suggests and show it to her again. Seeing her own ideas reflected in

your presentation is likely to get her to buy into your
agenda.

2. Let's say your technical talents and interests are greater
than your people skills. While it might be a good idea
to improve your ability to relate to colleagues and management more smoothly, you are probably wiser to pursue a technical career than one in management. To dis-
tinguish yourself, you might seek out specialties needed
by your company that no one else is willing or able to
deal with effectively.

3. Suppose you work for a medium-sized company and
you are competing against a corporate giant. So you
promote personal service and timely upgrades. If you
can offer a lower price, go for it.

   Conversely, you work for a major organization and
you're vying with a smaller competitor. In this case you
emphasize reputation, reliability, and perhaps a list of
satisfied clients.

   Calculating how to leverage your particular strengths
against the competition's weaknesses isn't always obvious and straightforward. Just keep the principle in
mind, and you'll discover increasing opportunities to
apply this effective strategy.

Synopsis
Cultivate your strengths to access opportunities and develop new strengths.

Conserve Your Energy
The right fight:
The right fight is the one you win without fighting.

Sun Tzu pointed out that, ultimately, victory is created by
your opponent through their mistakes and weaknesses, as
long as you are poised to take advantage of them. In modern terms, we might say that rather than bang your head

against a concrete wall, break the wall down with a sledge-hamer or, better yet, take a power tool to it. Use the best tools at your disposal. If tools are unavailable, seek a more vulnerable point of access. If you cannot find one, draw the opposition outside their defensive enclave to a more vulnerable terrain. Another tactic is to probe for structural weaknesses in the wall, areas where you can break though using a minimal amount of pounding or drilling.

In short, instead of attacking against your opponent's strength, draw them onto your turf, where they are weaker and you can bring your force to bear. Above all, don't spend your resources and energy unproductively (using a sledge hammer to open a bottle of wine would be wasteful in several aspects).

The wrong fight:
The wrong fight is the one that's not worth winning.

The wrong struggle is one where you have not found vulnerabilities that you are able to exploit. If an objective evaluation leaves victory uncertain or at too great a cost to justify the effort, back off: regroup, redefine your plan, or shift your attack to another target.

Maximizing your efforts efficiently is an ongoing learning process. Your objective is to achieve the greatest advantage with minimum exertion and waste, sort of like getting more miles per gallon in your car, and servicing it as often as needed (but not more often) to have it run for hundreds of thousands of miles. Eventually, though, there comes a time when certain costly parts are too worn down to enable you to drive the car safely. At this stage, a new vehicle may prove the wiser investment.

* * * * *

Sun Tzu also makes the point that true satisfaction is de-rived from remaining a step or two ahead of the opposition. Just about anyone can win with massively superior resources; true victory lies in prevailing through efficiency and ingenuity.

Synopsis
Most positions have their weaker points, or can be maneuvered onto less stable ground. Such is the lovely art of using less of what you've got to get more of what you want!

Summary
In the real world, no one is completely bulletproof. That being said, a woman must stand ready to defend herself against attacks from envious, jealous, and/or ruthless competitors, some of whom who will stop at nothing to trip you up.

At the same time, be vigilant for opportunities. Form a picture in your mind of where you want to be, cut back on obvious mistakes, and don't be afraid, or too modest, to represent your strengths. Positioning yourself for success doesn't have to come at the expense of friends and colleagues!

5. Performing
"... seek victory through the strategic configuration of power, not from reliance on people". – Sun Tzu

Chapter 5 is variously interpreted as performance through the use of power, energy, timing and momentum. It also features one of Sun Tzu's major strategies, that of using deception in order to confuse the enemy. He stresses the advantages of flexibility, maneuverability and swiftness.

Within this context, Sun Tzu speaks of applying timing to gain the advantage of momentum. He describes the use of "orthodox" (conventional) and "unorthodox" (unconventional) forces as counterparts in an endless cycle. Let's explore what he was telling us to find out what it might mean to you.

Employ Universal Timing

To go with the flow you must determine its pace and direction.

Universal timing is the sequence of natural forces, like the energy and predictability of tides and rivers. We employ the flow of water to create our metaphor, since many other forces (such as hurricanes and earthquakes) are much harder to envision or predict.

Practice

Practical examples of universal timing (moving in synch with the forces of nature) range from not having your umbrella turned inside out by gusts of wind to creating ergonomic buildings that take advantage of sunlight and soil. The Chinese, for example, have been successfully employing universal timing via living systems like feng-shui*, an age-old practical philosophy that endures into the current millennium.

Sun Tzu suggests that original thinking leads to innovative—and often unexpected—actions. Over time, of course, even the most unique innovations inevitably evolve into universally accepted strategies.

China completed construction on the massive Three Gorges Dam across the Yangtze River. They accomplished this by first diverting the river's flow around the site of the dam to be built, and later rerouting the water to its original course. In so doing, they coordinated human planning with universal timing.

---

* Feng-shui (literally, *wind-water*) is the ancient Chinese system incorporating the laws of Heaven (astronomy) and Earth (geography) to attain a better balance.

Today, diverting the course of rivers to build dams is common practice, but the first time it was conceived and used must have been considered a brilliant innovation. How can modern women profit from this principle?

Initially, by observing the flows around you: corporate policies, vested interests (harder to observe but worth the effort to uncover), power hierarchies and so on. Early on in your career, at least, you do yourself no favor by opposing these established flows. For while they may not necessarily be universal, they represent the prevailing and predictable forces that strongly influence your environment. In this case, bucking the flow is like swimming against the current: a challenge that opposes a virtual force of nature is unlikely to succeed.

Initiating Change

For the purpose of our discussions, we extend the metaphor into diverting the existing flow of male-dominated history into a more favorable course for women — not to exchange one form of dominance for another, but rather to stabilize the order into a more natural balance. Wouldn't a healthy equilibrium be more likely to endure far into the future than the lopsided state that precended it?

Alas, major changes of this magnitude are resisted by the incumbent power structure (mostly men) and thus face certain opposition. Here, however, opposing forces come into play. For one thing, women's issues are now in full bloom with a momentum of their own. Although not every woman appears quite ready to join the motorcade, you are certainly not alone in seeking equal treatment and opportunity. Second, women's rights represent a better balance than the male-dominated version, so we are, in effect, promoting a more natural flow than is currently the case.

Synopsis

Get in tune with yourself, and you will find the time to influence your environment. Changes that improve conditions for women favor all humanity.

Time your Moves

"Advance where the enemy does not expect it." – Sun Tzu

Personal timing differs from the universal in that it is determined by one's choice and judgment. But when you stop to consider good vs. bad timing, our reflections from the previous section carry over as naturally as, well, the tides nudged by the rising moon. Clearly, the most opportune formula for personal timing would synch up with universal principles before branching out. Which is to say, be mindful of the circumstances likely to affect your course of action before attempting to initiate a change or important decision. Be ready to move at a moment's notice.

Consider the Odds

The time to strike is when resistance is at its lowest.

Just as the best time to request extended responsibilities might present itself after delivering a successful project, it would be poor tactics to propose a new and costly initiative during budget cuts, ask for a favor when the boss is in a lousy mood, or start building a new road during rainy season. Interestingly, some research suggests that even weather conditions may color a person's receptiveness, so it might be wise to avoid asking for a big favor on a dismal, stormy day.

When you need someone else's buy-in, run the scenario through in slow motion: what are the plusses and the minuses? If they are approximately 50-50, then these are your probable odds of getting what you want; a greater negative probability should give you pause before pursuing your objective at the present time.

Can you do something to improve your chances? If so, then do that first to see what happens. Be flexible: if the odds still refuse to favor your plans, you are advised to shelve them until your probability of succeeding has improved.

What you don't want is to develop a pattern of failure where you and your ideas become associated with rejection.

Maximize Your Effort
Change happens both naturally and through imposition (unnaturally). The best change to be associated with is that which occurs in a natural and inevitable manner, where you can profit or exert an influence with relatively little effort. Maneuverability will allow to you take advantage of such opportunities.

Examples:
1. Edith's star has been rising in your department. Why not volunteer to contribute a few of your ideas to one of her projects before her likely promotion?
2. Your pc-oriented company is being acquired by a firm that favors Apple. This could be the perfect time to introduce a conversion to Apple within your department or company-wide.
3. You happen to know that Sally will be taking maternity leave toward he end of the month. Your boss might welcome your offer to temporarily take over one or two of her more visible responsibilities.

An example of unnatural (imposed) change might be, say, a departmental reorgainization that did not include consideration of the special skills, limitations, experience, and track record of the staff involved. A poorly managed organizational change usually results in disorganization, poor morale and loss of staff. Being associated with this kind of change is a situation devoutly to be avoided.

The Illusion of Change

Not all change is meaningful. Superficial (cosmetic) shake-ups may be introduced by managers and executives wanting to make their mark without upsetting the status quo. For instance, they might introduce new terminology (synonyms for existing terms), and move staff members to different offices and cubicles to make it look like things are different now that they're in charge.

A close analogy to Sun Tzu's war-based tactics in a civilian setting can be found in the deceptive practices of many politicians. Consider how those hoping to overthrow incumbents are forever advocating change. They promote this concept ostensibly to improve conditions for the voters, although the most important (if not only) change they're hoping for is to get elected. Ironically, these age-old strategies still prove effective because a voting population desperate for change is willing to believe the candidates' promises. This particular form of deception — saying one thing while meaning another — is what makes it work.

Synopsis

The best move is usually a matter of time, that is to say, when you are ready and your target (or competition) is susceptible.

Introduce the Unexpected

"He who excels at employing the unexpected is as resourceful as the Heavens." – Sun Tzu

Sun Tzu believed that initiative and momentum could be captured by doing the unexpected, a strategic precursor to thinking outside the box. He also advised us to "Use the orthodox to govern, and the unorthodox to run the military." What might this mean within an office setting?

Orthodox (conventional): Be straightforward: If you are a manager or administrator, treat your staff fairly and consistently. Let them know what is expected of them, and provide guidance, encouragement, objective feedback and (when appropriate) praise and reward.

Unorthodox (unconventional): Be shrewd. In a competitive situation, do the unexpected in order to reduce resistance at your intended point of attack. If, for example, A is your true objective, create the impression that you're really after B. (Of course, you'll want to let your boss know what you want, or she may reward you with the wrong prize.)

There are many ways of gently misleading colleagues and competitors to keep them from anticipating your intentions. Here are a few to consider:

1. If you want to talk to your boss after an argument with a colleague, go back to your desk and call her first to set up an appointment. Heading straight to her office is too obvious.

2. If you're trying to land an assignment that requires a specific talent (systems analysis, number crunching, writing skills, etc), prepare a sample of your expertise that relates to a different project, and then show it to your boss as if seeking guidance. "This is something I created at XYZ Co (or as a personal project, whatever). I was wondering if you could suggest a few improvements when you have the time."

   If she doesn't connect your experience and skill with the project you want, you could later follow up with something like, "What special skills are needed for the Rt-2 initiative?" Finally, if she still doesn't get it, you make it crystal clear, "Do you think my skills and background would be appropriate for the Rt-2 project?

3. Creating the impression that you're interested in a certain task or position could draw you into competition

with a strong competitor. Here's a thought: arrange to have a friendly discussion with that person, and then allow her or him to "talk you out of" competing with him/her. In this way, you offer to withdraw (from something you don't really want) in order to gain that person's support in an area that's really important to you.

Here again we call upon the Master's three-fold advice:
1. Be flexible in order to embrace a change of plans when called for.
2. Be maneuverable to change course with minimum inconvenience.
3. Be swift enough to catch the tide before it leaves without you.

Plan Your Strategy

Define where you want to go; otherwise, how will you know if you ever get there? Along the way, divulge only as much information as you want others to know. And remember, a secret shared is no longer a secret.

If you absolutely, positively need to share your personal goals and objectives, do so with a friend or family member, not someone at work. Why? Consider what may be to gain by blabbing (probably very little) as opposed to what you stand to lose (maybe a lot).

If you are a strong-willed, hard-working and productive individual, they're already aware of and watching you for any information they can glean. On the other hand, a quieter type is less likely to attract as much attention. Nevertheless, people are naturally curious and quite ready to think the worst. They seek your vulnerabilities, not necessarily to hurt you, but you might be surprised what some are willing to do in a competitive situation. So the less private information they have on you, the better off you are.

Apart from the obvious gossip gristle, competition can grow more intensive when people become aware of your objectives. It's human nature: as soon as one person appears to value something, others suddenly discover deep within their hearts a strong desire for the same thing. It's like the old saying we mentioned eariler: "People would throw away a lot more things if they were not afraid that someone else might pick them up."

Bear in mind too that posting information about yourself is extremely unlikely to further your interests, so why help others at your expense?

Innovate

The best-known form of innovation is a new or improved way to achieve a task or goal. If you manage to come up with a refreshing idea, think it through (and, if possible, test it out) before introducing it. What you don't want is to introduce a departure from the norm that fails, because that would lose you credibility.

Innovation isn't limited to doing things creatively and differently. In fact, it isn't something you can simply decide upon and do. For example, being watchful for potential opportunities, and then moving to position yourself to take advantage of them, is a great way to begin. Taking the initiative to get involved with new and vital tasks and projects is also innovative.

Anticipating imminent opportunities puts you in the best position to get your shoe (and maybe your entire foot) inside the door before the crowd. Of course, you'll need to prepare a reason for getting there, that is to say, a detailed plan of what you want, a rationale that justifies your getting it, and a reason or two suggesting why it might benefit the person makes the decisions. This requires you to be aware of what is going on around you – an ability to read the signs, and a link to unannounced changes and events (see Chapter 13 on networking).

Another innovative tactic that follows from awareness is the ability to stay a step or two ahead of the competition, such as rival colleagues and outsiders competing for the same market.

Avoid Predictable Mistakes

We talk elsewhere in this book about the applications and limits of practicing deception at your place of work. In the current context, we will limit our discussion to misleading certain competitors whom you have targeted for your own reasons. The guidelines are simple and straightforward.

1. Never lie: that's a reputation you don't want or need.
2. Never badmouth the competition: we are judged by what we say about others.

When you talk about people who are not present, only say what you want them to hear, because they will (negative comments travel twice as fast).

Express interest in your second choice (or something you don't really want), but don't let people know what you desire most.

Possible exception: a mentor or your boss (if he or she should ask).

Don't let people know that you don't like them: even nasty men and women imagine that they are well liked.

Leave ultra-personal, delicate, and potentially embarrassing details at home. Assume that anything that can be used against you probably will.

Synopsis

Predictability is nothing new. Let your good qualities (reliability, integrity, cooperation, etc.) be known, but keep your intentions to yourself.

Build Momentum

"The strategic power of using people wisely in warfare is like rolling boulders down a steep mountainside." – Sun Tzu

You can build momentum by initiating movement and maintaining it, or by re-directing something that is already moving.

Setting things in motion requires strength and energy; catching moving things demands exquisite timing. In both cases, maintaining their momentum needs persistence.

Everything being equal, propelling a vehicle forward from a standstill takes the most energy, building speed requires less energy, and maintaining a high speed demands the least of all. That's what momentum is all about. However much strength you might need to start a boulder rolling down a mountainside, once in motion it creates a powerful momentum of its own. In fact, it may be impossible to stop until it reaches the bottom.

How can you build a momentum that will serve your interests? By creating and maintaining balance, purpose and control. If this sounds like a contradiction, let us remind you of yin and yang, the opposite forces that together form the equilibrium Sun Tzu promotes throughout his writings.

Balance

The purpose of a ladder is to enable us to reach places that lie above our outstretched fingertips. We step up to the needed or available height, trusting it to bear our weight and maintain balance.

Ladders stand on two legs of approximately equal length; stools on three, and chairs on four. Although shorter ladders may open into a pair of two-legged sections, the taller extension ladders have only two legs. Why don't they build ladders with as many legs as chairs or tables? The answer, as you've surely guessed, is that more legs would make

them heavier and difficult to move about. Also, a narrow
base of even four legs (let's say the size of a chair) becomes
less stable as the ladder rises in height. The usual configuration is two legs on the ground with the top of the ladder
leaning against a wall, tree, or other solid surface.

However, none of this matters if the person on the ladder (you) is unbalanced, because you and the ladder would
then be out of synch. In this case, the strongest ladder in
the world could not keep you from falling. The point is that
wherever you may be – on terra firma, a ladder or your
place of work – your equilibrium depends upon your inner
sense of balance, not the props on which you may rely.

If you lead other people, they depend on you to maintain
a sense of balance to guide them wisely and productively.
Your schedules, distribution of labor and quality control
need to be coordinated — which is to say, balanced — or your
team will become less productive. Without this stability
you will find it difficult to maintain momentum in a positive direction.

In other words, the balance of your life (personal and
professional), and those who depend on you, begins and
ends with you.

Purpose and Control

Balance without purpose is like a scale with nothing to
weigh. Likewise, sitting quietly in the corner with a contended smile isn't going to launch you very far in your career. So don't get caught up using balance as an end unto
itself, because it is rather an approach to getting things
done. Or in this case, building a controlled and productive
momentum.

We're not suggesting that you lack meaningful goals,
only that you stand ready to choose, modify, and pursue
them as conditions and your inclinations change. Realistic (balanced) goals are those which are attainable within

a defined set of conditions and period of time. In order to achieve them, you must control the direction of the momentum you create.

By way of example, you certainly wouldn't expect to translate a 300-page cookbook into German in a fortnight, and the task would be totally unreasonable (unbalanced) if your command of German wasn't up to it.

Most of the deliverables for which you are responsible are less obvious and more nuanced. The point is that you need to pursue objectives in a controlled manner in order to build a positive momentum based on visible and meaningful progress, solid leadership, and effective management.

Synopsis
Maintain momentum, or it will stop you in your tracks; stay balanced, or you will spin in circles.

Create Your Own Reality
*"Begin the battle in a conventional manner
and win it unconventionally."* – Sun Tzu

This section relies on two of the concepts introduced earlier in this chapter, those of misdirection and timing. It's about knowing when, where and how (much) to control a situation to your advantage.

Franz Kafka once said, "In the fight between you and the world, back the world." What he meant was not that the world is necessarily right, but that if you continuously fight against it, you will lose.

Fortunately, you don't need to change the world in order to succeed, only to exert an influence on selected situations. Consider a large person with bulging muscles – would you agree to arm-wrestle him? Of course not; instead, you would cleverly divert the battle toward your area of strength (such as technical, communications, managerial, etc.).

For example, you might be better off opting to run a project than compete with an expert programmer/designer in her specialty. Your adversary's technical skills may be stronger than yours, but that doesn't mean she is a better project leader. Likewise, your product knowledge could make you a better choice for marketing manager than, say, a more experienced salesperson who is new to your company's product.

Any time you can convince decision makers of the value of your abilities (over someone else's) in a given situation, you have tweaked the parameters in your favor. This is what we mean by creating your own reality: it's a matter of influencing the environment just enough to promote your own agenda.

Negotiation
Be willing to give what matters less to gain what matters more.

Asian societies favor decisions and solutions where each party leaves with a feeling of having won at least an important concession. Although less often practiced on our side of the Pacific, this strategy has been successfully used by women throughout history. If this sounds like the win-win philosophy we talked about earlier, that's exactly what it is in practice. Compared to the win-lose principle, which is destined to gain more enemies than allies, you can clearly envision the advantage of allowing all the parties at the table to feel at least partially satisfied.

The Chinese and other Asians refer to this as saving face. We call it Psychology 101.

Preparation
Surprise is most effective in an orderly environment because it is, by definition, unexpected. Whereas the unpredictability that emerges from disorder is less of a surprise, since no one knew what to expect.

Frequent surprise, however, creates disorder, so spring your bombshells only when necessary, but not often enough for those around you to anticipate them. And do so from an environment that appears stable and structured. Which means that you need to well organized in order to:

1. set the stage for the occasional surprise, and
2. pace yourself with proper strategy and timing.

OK, let's step back for a moment to recall the heading of this section, which is to create your personal reality. As already explained, this metaphor refers only to a portion of your working environment, not the wider range which lies outside your ability to strongly influence. The purpose here is to subtly paint your picture in order to influence those around you share in your intended perception and support your place within that picture.

Your preparation is to clearly define what you want, identify and analyze the competition, create a list of qualities and reasons for people to support you, and sell it to the decision-makers. In other words, you create your world, and then invite them inside.

Synopsis
Set the stage on which you intend to act, and choreograph the performance.

Summary
The exquisite combination of universal timing, momentum and deception lies beyond the imagination of the majority of your competitors, unless they are familiar with the teachings of Sun Tzu. For one thing, universal (as opposed to traditional) timing is unknown to most westerners, as is the subtlety of deception (different from lying).

The same is true of Sun Tzu's concept of orthodox and unorthodox forces as complements to one another. That these are more typically Asian ways of thinking does not, however, exclude you from incorporating them into your western strategies and tactics.

Personal Profile — Nadezda Korn
"There are roads that are not followed." – Sun Tzu

Nadezda Korn emigrated to the US from Odessa, Ukraine. She is a highly resourceful financial information systems manager, mother of two boys, supportive wife and entrepreneur.

Flexibility is knowing when and how to move...or not.

I consider myself fortunate in always having had people to learn from. Most were on top of their game and left me with practical "how to" skills. A few others, at the opposite end of the spectrum, taught me "how not to." There was one special guy who managed to teach me both principles at once.

This manager would stake out a position that benefited his priorities best and maintain it as long as possible. Well, so do we all, to a degree, but he knew when to push forward and when to let go better than anyone I've ever encountered. His technique appeared to be inborn: an art, perhaps; definitely a gift.

A very clever fellow, he was somehow able to detect a potential flaw in his defenses well in advance of an opponent's notice. It might be no more than a hint, a key word, the shadow of a battering ram, and he could immediately recognize it as the beginning of an end. Suddenly and unexpectedly, he would eagerly embrace the opposing position and proceed to chart out a practical implementation thereof.

It was beautiful. In my mind's eye I think I actually saw people's heads spin. Instead of dragging out a fight doomed to failure, not to mention sucking up everyone's time and energy, he went on winning who knows how many tactical concessions. In addition, he often managed to unload the major portion of the work upon the jubilant victors.

I don't know if I can wholeheartedly recommend fending off the needs of other teams or switching sides like that, because this isn't in my nature. Still you have to give credit where it's due, and learn from the successful tactics of others. In this case, I've learned not to go on fighting the wrong fight.

Along with flexibility, I've learned the value of careful preparation.

"The art of war teaches us to rely not on the likelihood of the enemy's not coming, but on our own readiness to receive him; not on the chance of his not attacking, but rather on the fact that we have made our position unassailable." - Sun Tzu

When was the last time that things went exactly as planned? In my case, never!

We all enjoy a certain level of luck throughout our lives, to which we become accustomed and learn to plan accordingly. Fortunately for my business, I have no luck at all. In harmony with Murphy's law, things around me always seem to follow the worst case scenario.

So I plan by creating a plan B (in case plan A fails), and a plan C in case – well, you get the picture. Before a meeting about a contended objective, I go through "he said – she said" more times then when I have to coach a girlfriend on how to deal with her significant other. I write queue cards; I draw flow-charts. In all cases, I count on my adversaries to be at least as in-

ventive as I am. (I sometimes try to make them smarter, but for some reason it doesn't seem to work.)

Contrary to the popular catch phrase, there aren't
many suckers in the world at least, not within the
world of business. Which means you have to get
ready for some tough opponents. Which is why I try
my best to make my position unassailable.

6. Exploiting Weakness

"Now the army's disposition of force is like water." – Sun Tzu

In Chapter 6, Sun Tzu offers a comparison between illusion
and reality. The theme derives from Tao suggestions that
things which appear real (like power) may have little or no
substance; whereas a perceived weakness (vulnerability)
may prove stronger than you think.

Looking back to yin and yang, the Master reminds us
that reality and illusion (like strength and weakness) are
parts of the same whole. The message to women is that both
illusion and reality are ready tools for whatever image you
may wish to privately or publicly inhabit.

You may question the morality of exploiting someone's
weakness. Our view is that this depends on how and why
you do it, as well as on the outcome. We're not recommending
that you take someone's toys or job away from them, much
less cause them harm. But if your competition falls behind,
what's wrong with moving up to take their place on line?

Morality does not require you to step aside in order to
let others achieve their ambitions at your expense. We've
talked about win-win situations as preferable to the win-
lose variety. Yet circumstances are rarely so straightforward
as to create clear-cut pathways or distinctions. If others
covet what you have worked hard to gain, you are under
no moral obligation to hand it over to them.

Let's face it, as you climb the ladder of success, any number of less talented and motivated gals and guys are likely to believe that you got all the breaks they deserved. Deserved, not necessarily earned! This is an unavoidable reality of human nature.

The bottom line is that we aim toward honest and honorable goals (guided by principles like win-win) as best we can. It's reasonable to question one's own motives and methods along the way, but not to second-guess every step we gain, or wallow in guilt because not everybody measured up. Being a decent and successful woman doesn't guarantee that every helping hand you offer will be appreciated or rewarded. But that is not your problem, unless you make it so.

Carve Your Path

Water doesn't flow uphill (without a pump).

Water naturally takes the path of least resistance. In so doing it follows gravity, one of the most powerful forces known to nature. Trickles eventually become streams that feed rivers large and small. When you stop to think that our magnificent Grand Canyon was formed by water, you begin to recognize the domineering strength of its power and persistence.

You do not enjoy the luxury of centuries to forge your own path through mountains, but then, you don't need to since you're only facing human opposition. This is not to suggest that everyone is going to set obstacles in your path, but some will, intentionally or otherwise.

The application of this metaphor is consistent with Sun Tzu and the Tao which inspired his teachings: use your strength against their weaknesses; go with the flow, and do not waste your time and effort pitting your lesser strengths against the opposition's strong points. In other words, don't try to swim against the tide.

Examples:

1. If another woman has blazed a path within your organization, explore if it may be open to you as well.
2. If another path has historically been closed to women, try to find out why: does it appear to be the nature of the job, or might it be due to an executive's attitude?

This being said, attitudes can and do change as new realities begin to modify the landscape. One such change is rooted in the medicine profession, long a bastion of males. Judging by the growing percentages of female students in medical schools throughout the country, women appear poised to dominate this arena within a generation.

Synopsis

Some walls can be torn down, others circumvented. Highways are paved for travel, and a lot of choices lie open to you.

Define Your Style

You can't change who you are, but you can control the way you do things.

Style, as defined here, is about getting things done: the way you work and deal with colleagues, bosses and staff. It reflects your personal taste, preferences, and ideally your strengths and liabilities.

Lesser Strengths

Assess these qualities thoroughly and honestly. Give yourself credit for your assets and acknowledge your lesser strengths: no more talk about weaknesses, which can weigh you down like iron balls and dirty laundry. It's time to work at shoring up your lesser strengths while maximizing and applying your best qualities.

Remember too that strengths and lesser strengths are relative: one woman's lesser skills might appear substantial to another. For you, they're relative to the tasks you need to accomplish as compared to the abilities of your competitors.

What are strengths and lesser strengths? They are the objective (measurable) and subjective (judgment based) abilities people bring to the table. Measurable skill sets may be largely technical and analytical; subjective assets include management and leadership abilities; charisma, and political savvy. Be aware that these latter qualities are no less valuable—difficult to measure, perhaps, but essential to achieving career goals.

Perceived Strengths
One way to assess your abilities is to make a list of these and other skills relevant to your profession. You can divide them into plus/neutral/minus columns, or rank them from one to ten in order of perceived strength. Then (and this is important because women commonly undervalue their abilities) have a few trusted friends and colleagues rank them (give them a clean sheet without your numbers). You might be surprised at the results.

Now re-rank these qualities and review the list as if your career depended on it. Note major discrepancies between your rankings and those offered by your friends and colleagues, especially if they tend to agree with one another but disagree with you. These can illuminate a few blind spots where you under or over value your abilities. The objective is to develop a clear picture of what you've got going for you, and which lesser strengths may need to be improved.

When one of your authors was in grade school, his teachers told him to focus on his strengths and not worry too much about his weaknesses as long as they were good enough to pass exams. The other author, raised according to Chinese tradition, was taught to achieve a better balance by elevating her lesser strengths into assets.

Sadly, our school days cannot be revised or relived, but the future is still waiting to be recorded.

Armed with a clear idea of your relative strengths, you're ready to decide which approach might work best for you: are you verbal and outgoing or reserved; comfortable in groups or better one-on-one; impetuous or patient; flexible or rigid; decisive or prone to procrastination? The style you assume must be comfortable yet challenging, making use of your strengths while encouraging you to build up your lesser strengths. Package your style into a brand with which you want to be identified. Seek a balanced (yin and yang) approach to work, relationships and career in a natural and consistent manner, just as water flows.

Synopsis
Discover who you are, and then let others know as much as may be suitable to your objectives.

Take Initiatives
Initiatives are voluntary risks; innovations are things done differently.

Initiatives
Since men enjoy a reputation for taking risks, initiatives are less expected of a woman. So much the better, as this may enable you to stand above the crowd. Note, however, that the risks you initiate will be scrutinized and second-guessed to a greater degree than those of your male counterparts. So choose them carefully, in accordance with the style you have adopted and your probability of success.

This probability we keep mentioning is not a question of precise percentages like those established by gambling casinos. You've got to rely on your analysis of the scope, requirements, available resources, time frame, and support behind any endeavor, plus the value of succeeding. We're talking about risk vs. reward: after all, there is no more reason to initiate a heavy task that will not be appreciated than to fight against heavy odds. Clearly, pouring water into a dark hole is no more rewarding than attempting to force it uphill.

Example:
If you intend to approach a task in an entirely different manner than may be expected, explain (in writing and verbally) the manner and reason for your innovation. If you cannot justify it or gain support, reevaluate your plan.

Innovations
There's plenty new under the sun—aren't you unique?

Innovation can be a brand new approach to getting something done, or applying a known technique to a novel use. When an innovation works, glory to the innovator. If not, you've hopefully learned something worthwhile during the process that will make the next attempt more reachable.

Doing things differently is, by definition, a departure from the norm. Since most people are routine-oriented, they tend to resist any change that nudges them outside their comfort zone. Such is the nature of resistance you can expect when you attempt to innovate: the greater the innovation, the stronger the resistance. Multiply this by the gender factor, as many men strongly oppose re-direction by women, and you have identified the bulk of opposition.

Let's not overlook other potential factors, like colleagues or managers who have tried unsuccessfully to resolve the problem you're confronting and may have a negative stake against being outshined by a woman. Of course, convincing one or two of them to collaborate and share a measure of the glory (if successful) is a tool in your creativity bag, if you're so inclined.

Best of all are innovations borrowed from your past experience which seem suited to the present situation. No one, other than your boss, perhaps, needs to know that your innovation is essentially a retread. Unique or not, an effective innovation can be a career-enhancer for sure.

Innovation can also be applied to reducing opposition, as by co-opting major pockets of resistance in advance. One tactic we've already suggested is to invite participation in your approach. Another is to ask for suggestions and advice on minor aspects of the innovation. Then, by crediting certain individuals for their invaluable contributions, you may have neutralized their opposition, if not gained their support.

Synopsis
Take reasonable risks that offer commensurate gain. Begin with the conventional method and then seek a better way.

Leverage Strengths
"One who excels at warfare compels men and is not compelled by other men." – Sun Tzu

As water follows its natural course, mountains of earth and solid rock irresistibly give way, widening pathways and eroding constraints. Thus does persistence gain momentum.

In your career, small successes will add up to greater opportunities and responsibilities. You rise within a hierarchy where glass ceilings may replace more overt forms of gender bias; you attain power within subtle restraints; you begin to leverage your growing strengths in several directions:

Upward
Unless another woman has paved the way to the executive suite, you are now on rarified ground. Evaluate the turf in

front of you (Chapter 11) and proceed accordingly. Tradition – the good old boys club, along with the entrenched attitudes they probably don't even recognize – may stand in your way. This is where your charm and feminine instincts can serve you well. As a rule of thumb, your strategy to infiltrate this turf is to present yourself as:

1. an ally to at least one or two key players,
2. a benefit to the team, and
3. absolutely no threat to initiate major changes to the comfort zone they have established.

Once you're in the door and firmly situated, you can spread your elbows just enough to make yourself a bit more comfortable. Performed with subtle skill, chances are the guys won't notice the differences.

Outward

The higher you scale the corporate pyramid, the more intense the competition for fewer choice positions. Remember the game where everyone rotates around a circle of diminishing chairs? Someone always gets left standing without a place to sit and is thus eliminated. The contest continues until a single person possesses the one remaining chair and is declared the winner.

Corporate positions are much like this game, complicated by moving chairs and brand new players. Executives may switch positions, and someone may be brought in from outside to occupy an empty office. The positions you and your peers are competing for may not be available to any of you.

All the while, a strong sense of competition prevails and not everyone plays nicely. Do not be surprised if your competitors bring rough and ruthless tactics into the mix in order to outflank you. And don't be deluded into thinking that merit necessarily perseveres over politics, because the opposite is more often true. So decide if you want to compete at this level, enter the arena with awareness of the turf, and gear up for a version of competition that more closely approaches Sun Tzu's analysis of war. Hold the thought that if they (the guys) can do it their way, you can do it yours.

Downward

Value and nurture the people who report to you. Be fair and consistent; build morale and loyalty, and maintain an environment where it is worth your staff's while to give you their best efforts. Support the managers and leaders who depend on you without getting involved in unnecessary detail. No matter how intense the politics where you are sitting, remember what got you there in the first place.

Synopsis

Build on a base of strength in all three directions, and use your strengths as much as needed. Remember, you can blow out a candle, but not a forest fire.

Minimize Your Lesser Strengths

No woman is an island.

There is no wall that cannot be breached, no team is immune to defeat, and no individual of either gender can do everything alone. The reason, of course, is that we all have vulnerabilities—blind spots where we can be attacked and damaged. Shoring up your lesser strengths may only go so far, because no one can be expert in all things.

There are several ways to protect, to a degree, against exploitation of your vulnerabilities. These methods are by no means mutually exclusive, so feel free to mix and match as may suit your style and needs.

Unite

Some of your competition is as clever and resourceful as you. Such a colleague may be open to the possibility of a strategic alliance, where working together can achieve more than twice what one might accomplish on her own.

There are plusses and minuses to selecting a male or female partner, so that is not your initial consideration. A functional collaboration must be based on:

1. trust: You don't want to team up with an unreliable person who may abandon you in mid-flight;
2. complementary skills and talents: While certain of your strengths may overlap, the vulnerabilities of each should be covered by the other's strengths; otherwise, what you've got is little more than a mutual admiration society.
3. compatibility: In virtually all human interactions, an ability to get along is essential (incompatibility is a deal-buster).

Evade

You may be able to carve yourself a niche in which your lesser skills don't come into play. For example, a project manager is unlikely to possess or need the technical skills of the analysts she manages. However, many technical specialists dead-end themselves by neglecting the managerial and leadership abilities they'll need in order to succeed as managers and executives.

Upgrade

Why not convert a few of your lesser strengths into assets? Most large corporations offer classes or tuition reimbursement, and numerous on-line courses are available to a woman willing to make the effort. It's about investing in yourself.

Sidestep

Should you find yourself in a situation where your responsibilities outweigh your strengths, consider a career change in a direction that better suits your interests and abilities. More easily suggested than pursued, to be sure, but for ev-

ery month you spend retooling, you're likely to enjoy a year
or two of satisfaction in a more rewarding career. Consider
as well that if you are uncomfortable and insecure where
you are, you're probably not the only one who's noticed.

Suggestion
Make a list of your strengths and lesser strengths to the
best of your knowledge. Next validate and prioritize these
items, and then discuss them with a trusted friend or colleague (why not both?). The point is to develop an accurate
idea of what your strengths and lesser strengths really are.

While you're doing this, please remember that your
strengths and weaknesses are complementary (yin/yang)
elements to be kept in balance. Accept the plusses and the
minuses as part of your toolbox to rely upon and improve.

Synopsis
Use what works for you, and find ways to compensate for
what doesn't.

Summary
The fine lines between illusion and reality can also be described as isolated levels of perception. The question is,
how can perception (belief) affect the physical world?

Within the current context, we are focused upon influencing the behavior of our opponents, and to a lesser degree, that of colleagues and decision-makers.

Like beauty, that which appears real (for instance, a perceived problem or threat) has substance in the mind of the
believer. Certainly, worrying about a threat can cause a person to grow apprehensive and affect their health. Whereas
the ability not to be concerned about the same supposed
danger produces an entirely different effect. So until the issue has been verified or disproved, its physical reality may
cause less damage than the perception.

Now, if the physical threat (let's say a computer virus) does not exist, does that make it any less real to those who are afraid of it? Conversely, is a real virus any less a threat if people refuse to believe that it exists?

The point to this little exercise is to emphasize the power of perception, and the possibilities for deception that it enables. This applies to weaknesses and strengths, opening and closing doors, and representing any aspect of yourself in any way you wish.

Here is yet another example of the intricate relationships between apparent opposites, like yin and yang.

7. Positioning

"War is based upon deception." – Sun Tzu

The focus in this chapter is on maneuvering to occupy a favorable position before the opposition gets there. Sun Tzu's tactic for accomplishing this vital principle is based upon learning as much as possible about your competition while revealing nothing of value. In fact, the author recommends misleading them by camouflaging your intentions plans and strategies. The objective is to select and then reach your destination (position) first and without having to endure a power struggle.

Sun Tzu moved armies consisting of thousands of men, horses, carts, munitions, food, and other supplies into strategically important positions. Your task is far less complex but no less important. Let's see how the Master's wisdom can help you to position yourself for success.

Project Your Image
Reveal with Purpose
Let the people where you work know only what you want them to know about you. In order to accomplish this, you

need to focus on projecting a positive image of yourself that
suits your personality and objectives.

Yet you may, in unguarded moments, reveal more about
yourself than you might wish. Without belaboring the
point, here are a few pointers you may find useful:

1. Try to be supportive, upbeat and positive. If asked
   about someone's obviously poor performance, be objective rather than bashing them.
2. Never exaggerate your abilities or experience. Better to
   express an interest in learning more about something
   than to try to hide your ignorance.
3. Associate yourself with successful projects. Better to
   participate in a positive result than to lead a less successful effort.
4. Give and share credit with others who have earned it.
5. Don't pass the blame to someone else, even if they deserve
   it. There are more subtle ways of making the truth known.

Advertise
Another image-builder is self-advertising. Not the blatant
"I'm so great" form of egomania, but a subtle style much
better suited to women.

First of all, get it into your head that advertising works!
After all, the media has been using it for decades to sell
inferior products at premium prices. How does it work? A
major reason is the power of suggestion: communicate your
message over and over again, and it eventually sinks in to
the subconscious. Of course, the same pitch doesn't influence everyone in the same manner, but overall, advertising
has proved its effectiveness.

How can you cash in on this principle without over-selling
yourself or compromising your values? First of all, you need
to decide how you want those around you to perceive you.

As far as your expertise and professionalism are concerned, your colleagues evaluate you according to what

they observe. Obviously, you cannot expect to sell them on the quality of your work, since this is already visible. Your message, then, will be oriented toward intangible qualities, like your character and integrity.

Example:
Other person: "Josh was saying that you aren't very easy to work with. I've heard he can be pretty stubborn himself."
Response 1: "Really? Well, I don't like retaliating."
Response 2: "Personally, I prefer not to add fuel to someone else's fire."
Response 3: "Hopefully you and I will have an opportunity to work together. Then you can decide for yourself."

Bottom line: Whichever of these approaches suits your style (or something entirely different), this tactic can be used to shape, or at least influence, your image in the workplace. Just be sure that your message is consistent with your manner and behavior.

Synopsis
To control your public image, let them know only what you want them to know.

Guard Your Vulnerabilities
Jealousy
Given the extra difficulties women encounter in predominantly male-dominated environments, you'd think they might band together to help one-another. While female alliances do occur, not all women view each other as sisters. Surely you have noticed that the source of much resentment over your talents and accomplishments is other women.

Lois S., a medical doctor serving her residence in a Chicago hospital, told us, "The (female) nurses were the hardest to deal with. They were always second-guessing me to show me how much more they knew than me. And while some of them were very experienced, they didn't treat the male residents this way."

Women, far more often than men, scrutinize the way their female colleagues dress (especially their shoes), how they wear their hair, their cosmetics, handbags, and how well their clothing fits. Men notice some of these things but are more interested in the overall effect (attractive, approachable, etc.).

Females are also more difficult to manipulate because they're on to all the wiles that work on most men. Worst of all, they are, by and large, more threatened by an attractive, intelligent, competent and friendly female whom they may view as a threat to their status in the office.

For the record, we're not suggesting that all or even a majority of women behave this way, only that far more of this instant resentment comes from other women than from men. What to do?

1. Take a stand! Show everyone the persona (face) you want to project, clearly and unambiguously. Avoid female colleagues whose eyes aim negativity at you, and gravitate toward those who seem friendly.

2. If another woman is treating you badly (saying nasty things about you and generally trying to put you down), consider some version of: "I don't know what you've got against me, since you don't know me, and I've never said a word against you, but I'd suggest you put a stop to it right now. Because if you're really looking for an enemy, I can kick back at least as hard as you can."

   Whatever you say, say it in a soft and even tone, without a hint of anger. If you can manage to smile, so much the better. If she tries to brush you off with the old "Why, honey, I'm sure I have no idea what you're talking about," all you need to do is smile, look her straight in the eye, and leave without getting into an argument.

As difficult and unpleasant as this kind of encounter can be, the result of allowing the problem to continue unchallenged may become a nightmare. So nip it in the bud, if possible—you'll sleep better.

What if you do all this and it still doesn't work? Then you may want to begin documenting the troubles you're having. Find someone who is willing to back you up, and then present your case to your boss or (if a different person) hers. Frame the issue as causing dissention in the office, and be sure to point out that you have done nothing to feed the flame. "I'm not looking for any kind of retribution," you might emphasize, "I just want it stopped."

3. If a woman seems cold to you, take a shot at warming the interaction.

Example:

"Hi. As you know I've only been here a short while, and I've heard that you really know your way around the office. I wonder if we could have lunch together when you have the time – maybe you could give me a few tips about the do's and don'ts."

Sometimes this approach will work, sometimes not. She might respond,

"Sorry, but I usually eat at my desk."

In this case, smile pleasantly.

"Of course, I understand," and back off. But continue to be friendly and respectful—she might warm up to you eventually and is unlikely to become an enemy.

Be wary of the women who do befriend you until you know them better. Above all, leave any but the most mundane details of your personal life out of the office.

Criticism

The best way to deal with criticism is to consider it. A valid critique, whether offered with positive (to help you improve) or negative (to put you down) intent, is still useful. Just as you would prefer to learn from your mistakes instead of repeating them, use any negatives to help you improve.

How can you decide if a criticism is worth pursuing? Several ways:

1. Have you heard it from someone else?
2. Is the source someone who appears inclined to help you?
3. Does their observation seem reasonable when you think about it later?
4. Does it make you so uncomfortable you don't want to think about it?

If any one of the above is true, there's no harm at least considering the point and maybe asking someone else about it; if two of these conditions ring a bell, you should strongly consider it; and if three or four of them are accurate, you have probably identified an issue in need of correction.

Note that your response to criticism will determine whether well-intentioned people will be willing to offer you their candid observations in the future. So be sure to thank anyone who does so with something like,

"Thanks, Shirley, I appreciate your telling me."

If someone's criticism turns out to be potentially valid, you might get back to them with,

"About that pointer you gave me the other day, I'm working on improving... (a better approach, etc)."

Even if the criticism appears frivolous or malicious, continue to smile without letting the person know you have rejected it. If they should ask you about it, you can always say something like,

"I am definitely thinking about it."

Bottom line: You don't need to jump through hoops every time someone has something critical to offer you, nor do you want to develop a reputation for rejecting criticism.

Synopsis

Protect yourself where you are most vulnerable, and plan to deal with negativity. No one's ride is smooth from here all the way to wherever they are going.

Undermine the Competition

Not all of Sun Tzu's military tactics fully extend to modern business practices. Obviously, we aren't going to recommend damaging your colleagues' reputations or careers, only to out-position them toward specific goals. That being said, it is to your advantage to learn as much as possible about your major opponents, such as who else might be trying for a role you have targeted.

It would be ill advised to pump them for information by pretending not to be on the same trail. Such deception might work once, but it would earn you their enmity and distrust thereafter. A better tack might be to ask, "I wonder, beside Mel and me, of course, who is most interested in GYQ, RPM and ##."

This level of curiosity would not be taken as unusual within a department or corporation, as rumors often provide some of the more revealing conversation. Our point is that if you are able to identify your most serious competition for RPM, you might be able to tweak your proposal to emphasize your strengths against their weaknesses. When the truth eventually comes out, they might be surprised or annoyed with your strategy and tactics, but no one could accuse you of being deceitful or underhanded. In the final analysis, not all competitive situations enable win-win conclusions.

If you win the RPM project, you might approach one of your competitors and invite him to work on (but not equally share) the project with you. This would especially make sense if he had a valuable skill to offer. "I would really appreciate collaborating with you on the bottleneck problem in RPM, in which I understand you have strong background." If accepted, you could credit him for his assistance in a specific area without diluting your responsibility for the entire project.

On the other hand, if you are competing with an outside organization's products of services, you might consider taking a stronger stand. But once again we would discourage any false or truly misleading claims, because they are more likely to hurt you in the long run.

Synopsis
Competing on your terms gives you home-field advantage.

Protect Your Health
"An army that avoids the hundred illnesses
is said to be certain of victory." – Sun Tzu

Your brain is the force behind your judgment, motivation and well-being. For better or for worse, it lives inside your body. Whatever challenges you face, remember always that a healthy mind deserves a healthy place to live.

While this advice applies to men as well, we have found that women are more prone to sacrificing their own well-being as they juggle careers with family responsibilities. The problems involved in raising children are known to mothers everywhere. It also seems that husband and wife relationships weigh more heavily upon women. These tendencies add pressure to the workplace, where women are expected to compete within environments that are often weighed against them.

Sure, it's easy for us to recommend proper diet, exercise, and healthful living habits while you're out there struggling to survive. Yet we emphasize the point, because poor health can become one barrier too many.

Balance

Aside from the obvious, like not putting poisons (cigarette smoke, excessive alcohol and all the rest) into your body, be conscious of the food you eat and how it affects your weight and energy. Too much of this and not enough of that will upset the balance of your life flow (remember yin and yang). If you're only burning brainpower and leaving the remainder of your body to fend for itself, you risk the onset of physical issues like poor sleep, joint and muscle soreness and low energy. Can illness and depression be far behind?

Don't bother answering that you don't have the time to exercise, because few of us do. At least half the people you find working out in health clubs have no more free time than you, and yet there they are!

The point is that exercise in a gym, outdoor jogging, or using a machine at home all require discipline. You've got to schedule a time to exercise, just like any other important activity. It's a meeting of your mind and body, a precious communion that belongs to you and no one else. Treat it as your private time away from the rest of the world, and (short of an emergency) don't allow anything or anyone to interrupt it. The good news is that once you form the habit of exercising it gets easier to maintain, and you'll feel better physically and mentally.

Health Problems

"When their strength has been exhausted and their wealth depleted, then they will be empty." – Sun Tzu

People become as easily addicted to sadness as to joy. This is because most of us are creatures of habit and resistant to change. The expression "Better the enemy you know" is generally poor advice. Rather, turn the enemy into an ally, render them harmless, or hang out in a friendlier circle.

Physical illnesses are best treated by health care professionals, but these specialists can only help you if you visit them. Caring for the inevitable illnesses and injuries of children blinds many women to the importance of tending to that nagging soreness in the neck or back, or maybe a persistent cough. Hey, if it's something for which you would take your son or daughter to the doctor, do the same for yourself. You know that conditions usually worsen when untreated, so grant yourself the same respect you give your loved ones.

Boredom, chronic fatigue and depression commonly derive from monotony, lack of motivation, and of course stress. These are more subtle foes to your wellbeing, but no less serious and important to address.

Stress

The most overlooked cause of career-related illness is stress. This subtle and yet powerful condition is caused by physical and mental changes heavily imposed on us. Its subtlety stems from two factors: first, the signs of stress begin almost imperceptibly; the second factor is that its symptoms can serve as a helpful warning mechanism to alert us to the possibility of an impending threat.

Stress often results from indecision about how to deal with a threat or serious problem. It is largely caused by the perception of the threat. Perception is the initial flag, because the anxiety is mental: whatever the crisis, it's only stressful if you are worried about it. The second indicator is indecision, because if you are unable to decide upon a course of action, your anxiety is unlikely to diminish.

In other words, stress is basically your mental reaction to the problems you recognize but cannot decide how to resolve. It can be triggered by issues at home or office, or simply the possibility of a threat. Worrying about the health of a loved one, uncertainties at work, financial difficulties and emotional concerns are all potential harbingers of stress. Whether valid or imaginary, the effects are similar.

In reality, a small dose of stress may actually help to you prepare for an important exam or presentation. Larger and prolonged amounts, however, are more likely to put you in a bad place.

The signs, visible to those around you, are fatigue, irritability, difficulty sleeping, negativism, depression, deteriorating health, lowered performance, and any other uncharacteristic behavior patterns. People who know you are more likely to notice these changes; others may believe that this is who you really are.

To put it bluntly, stress is something you can live without. However, since it hardly ever disappears on its own, you are advised to deal with stress as soon as you recognize its signs. In fact, the earlier you jump in, the less difficult it may be to confront whatever happens to be causing it. To this end we offer the following recommendations:

1. Try to recognize even the subtle signs of stress in your moods and behavior.
2. Jot down the issues that seem to concern you, and divide them into a) things you can easily resolve; b) things that are difficult but possible to resolve; and c) things you don't know how to resolve.

Set about fixing a) as soon as possible; start working on the b) issues with patience and persistence; and seek help from a good friend, trusted colleague, or family member in order to approach the c) problems.

Believe it, the process of recognizing your troubles and working on (even a few of) them will help improve your spirits and energy flow. This is the best way to reverse a downward spiral into an upward flow!

Synopsis
Make the time and effort to attend to your own well-being. You can't be there for others if you fail to help yourself.

Maneuver to Advantage

"Never fight a losing battle." – Sun Tzu

Sun Tzu cautions us that successful maneuvering to attain strategic positions is not an easy task. The major difficulties lie in "making the most devious road the most direct," and the "artifice of deviation."

One of the most effective ways to spin your competition off track is to tell them a truth that they will not believe.

Examples:
1. "I couldn't sleep all night worrying about your counter-proposal (yawn)."
2. "I have no idea what you're referring to (knowing look)."

Here's how you might apply this principle into a competitive strategy:

Suppose several projects are up for grabs in your office. GYQ is the most popular project, but you're gunning for RPM, which better suits your interests and abilities. A third and lesser project is also in the offing – we'll call it #3.

Mel, an experienced manager, is making a big push for GYQ. He appears to have an inside track, but has also expressed interest in RPM as a possible fallback. Other members in the department are maneuvering for RPM and #3, but you're not entirely sure who they are or how strong their petitions might be.

You have no interest in GYQ, much less in opposing Mel head-to-head, and you don't want anyone to know about your interest in RPM until the time is right. So you toss out a waffle ball to hide your intentions when the topic is being discussed.

"Actually, I have no interest whatsoever in GYQ," you state with a confident smile. This predictably creates the impression that you really are interested in GYQ.

Mel, of course, hears about this and confronts you: "I hear you are my main opposition for GYQ."

You respond, "What makes you think that?"

The bait is swallowed and the hook is set. But instead of reeling in the fish, you leave it out there in the river fighting an imaginary adversary. Meanwhile, none of your potential competitors for RPM is giving you a second thought, enabling you to quietly position yourself for the prize you really want.

You create a solid proposal, let your boss know about your focus on RPM, and ask her to keep it in confidence as long as possible. You might also request her advice as to how you might polish your proposal to give you a leg up on the competition. She may refuse this request to avoid showing favoritism, but the possibility of slipping into an inside track is worth the risk. She might even be willing to discuss the project with you, offering valuable insights. In any event, your interest, enthusiasm and positive approach are unlikely to offend a reasonable manager. If anything, these qualities could elevate you above the competition before they have even arrived.

You have misled (without lying) the competition and positioned yourself with the decision-maker. Now it's time to justify your ambitions with a great proposal.

Synopsis
Don't jump on ice if you can't skate. Instead, move the game onto familiar turf where you have already learned to cruise.

Summary
Positioning, a vital sports and military tactic, is vastly overlooked as a career concern throughout the western world. It's not as if the principles were unknown, but they are rarely emphasized or used to gain advantage.

Terms like misleading and camouflage are, perhaps, distasteful to some women who prefer to pursue open and honest relationships. However, the manner in which Sun Tzu recommends these practices should pose little if any conflict. He is not, after all, suggesting that you lie or cheat —emphatically not! He only advises that it isn't prudent, and may serve to your disadvantage, to reveal all.

Is it necessary to let everyone in your office know about your short, medium and long-term ambitions? Whom you respect most and least among your colleagues? Your fears and anxieties? Your perceived weaknesses? Of course not. Clearly, the latent disadvantages would far outweigh potential benefits.

Face it—interpersonal relationships are a delicate balance of truth and social decorum. You would no more tell your boss that her skirt made her look plump than point out a colleague's receding hairline. Whether tact or strategy, is it honest to hurt somebody's feelings and reveal your vulnerabilities, or just poor judgment?

8. Navigating

"...create an impregnable position." – Sun Tzu

This chapter centers on the essential conditions that dictate successful military strategies as reflected throughout *The Art of War*. They address which paths to follow or avoid; when to move forward, hold your position or retreat; and why certain rules and instructions should be ignored.

Sun Tzu also cautions us about certain risky characteristics which leaders (and by extension, you) are advised to avoid. These include extreme pessimism or optimism, uncontrolled emotion, obsession, and excessive sentiment.

As elsewhere, the underlying messages are preparation, anticipation and flexibility. We have organized these into practical, yet flexible principles for the working woman.

Develop Credibility

Confidence, performance and consistency are convincing.

As a manager, leader and/or employee, professional credibility is no less essential to your career than wings to a migrating swan. Managers and colleagues will entrust you with opportunities, responsibilities and confidence only to the degree that they believe in your competence and integrity. Think of credibility as your brand name, a reputation that can either clear the way before you or clutter your path with obstacles.

The tried and true method to create credibility is to develop a reputation for succeeding without damaging your company, co-workers or yourself.

Easier said than done? Not really, if you are a reasonably competent woman and know how to spot and avoid predictable snares—like a fox that, no matter how hungry, avoids a baited trap which appears dangerous.

Given Sun Tzu's advice about preparation, patience and judgment, the objective here is to recognize circumstances and situations with a high probability of success. In other words:

1. observe and consider the known circumstances likely to affect success or failure;
2. be ready to exploit the advantages you discover;
3. remain flexible to adopting alterative tactics and strategies when appropriate.

    These simple concepts will help you to develop confidence in your decisions and abilities.

Sun Tzu also identifies negative leadership qualities, in particular:

1. Extreme pessimism or optimism: A negative frame of mind will pull down those who are depending on you as surely as a balloon devoid of helium. No matter how difficult things may be, expecting to fail can be the coup de grâce to your plans and projects.

   Conversely, unbridled optimism has been known to blind people to the effort and details needed to be successful.

   As always, it's best to seek a yin/yang approach, using positive and upbeat energy to commit yourself to the effort needed. In other words, can do is only useful if you make the effort.

2. Uncontrolled emotion: If asked to choose between a hothead and a crybaby, which would it be?

   If you answered, "Neither," we're on the same page. Surely you have noticed that screamers and weepers are equally unbalanced and unreliable. Likewise blamers and those who feel that everything that goes wrong is their fault. Blame, whether directed outward or inward, simply isn't productive.

   Whether at war, on the operating table or an office, a calm and rational approach inspires confidence and promotes clear thinking and action.

3. Obsession: The need to be perfect will never be fulfilled, because no one is. It follows that an obsession with performing spotless work is vulnerable to the inevitable mistakes we all make.

   We're certainly not suggesting that lower standards and mediocre performance are acceptable, much less any other extreme position. We're talking about reasonable expectations and professional equilibrium. Balance. Yin and yang.

4. Excessive sentiment: Sun Tzu's criticism of rulers who

love their subjects overmuch is fairly obvious within a military setting, where soldiers are often asked to risk their lives. In work and personal settings the message is more subtle, but similarly valid.

Could you, by way of example, assign demanding tasks to someone you felt a need to protect, criticize poor work to a person whose feelings you didn't want to hurt, or compete against someone you couldn't bring yourself to challenge?

Forming friendships, working partnerships, and other mutually advantageous relationships can be worthwhile, even vital in the right circumstances; just don't let your emotions run away with your judgment and common sense.

Synopsis
Credibility is your trademark, the basis of your reputation. Earn it, protect it, and bank on it.

Weigh Circumstances
"...a wise leader weighs the sum of gain and loss." – Sun Tzu

Sun Tzu refers more than once to the kinds of terrain a leader may encounter, and how to react accordingly. There are places not to enter or linger, forbidden pathways, ill-conceived instructions, and conditions which offer poor odds for success.

Elsewhere in this book we encounter situations that seem doomed to failure: unreasonable timelines, insufficient resources, poorly defined objectives, and unclear expectations.

Perhaps a few military examples would be appropriate here:
1. Would you send your troops into a forest where the enemy might be hiding?
   Probably not without first sending out a scouting party.

2. Would you enter a battle without proper ammunition?
   Not on your life!
3. Would you try to swim across a river with a strong current wearing a knapsack and 80 pounds of equipment?
   It's unlikely even Michael Phelps would try it wearing his gold medals.

Then why accept unfeasible assignments and situations where the odds are stacked against you higher than your line of vision? For example, suppose you were asked to join a marketing team that had just lost two senior staff members. A warning flag should pop up in your mind's eye immediately: am I expected to pick up the load for both of them, or has a specific task been defined that I can reasonably handle?

Women are more prone to accept commitments based on trust, ceding to the judgment and good intentions of their boss. From now on, however, you are going to actively participate in evaluating your assignments, right? Right on!

Synopsis
Be prepared to walk away from a bad deal. Don't let vanity, fear, or wishful thinking cloud your judgment.

Be Flexible
Flexibility is the capacity to bend without breaking.

Knowing when to change course is a sign of preparation and flexibility; the opposite is being stubborn and getting caught flatfooted.

Momentum can work to your advantage by helping to conserve energy, or against your interests by resisting change. It may be habit-forming, comfortable, and hard to resist. But when you begin to lose control of your pace and direction, it's time to pull back and reconsider options. Especially if you find yourself out in the middle of nowhere or faced with a series of dead-ends.

Bottom line: Momentum is only useful when it serves your purpose. Otherwise it may be rolling you.

Flexibility is also a two-edged sword. On the one hand, a degree of flexibility is absolutely essential to keep you from leaping off cliffs or into unfathomable depths. Being ready to shift gears when the circumstances suggest a new speed or direction is a wonderful ability, one that requires strength, courage and confidence. However, being too ready to curtail or redirect your efforts may signify just the opposite: a lack of confidence and fortitude.

As a woman, you may be more subject to criticism than your male counterparts. Judgment to the rescue, for only a clear and ongoing assessment of your circumstances can prepare you for a wise decision: stay the course, alter your direction, or back away from a losing situation. Your best solution lies in objectively analyzing the nature of the difficulties, and the amount of effort and resources needed to resolve them.

In a no-win situation, Sun Tzu recommends that you try to improve conditions in your favor, alter your plan, or withdraw. He sensibly advises against taking a desperate risk where failure is the likely outcome.

1. Staying the course: Challenging tasks are rarely cakewalks. They usually require effort, planning, determination, and the flexibility to make changes along the way. This is where you show yourself and everyone watching whether you've got what it takes to deliver.

2. Changing direction: A major change of emphasis may be called for in the face of severe budget cuts, downsizing or adjusted priorities. Your ability to recognize that the original project needs to accommodate major modifications and suggest acceptable alternatives may save the project and elevate your image all the way up to the executive suite!

3. Backing off: When a fisherman's line is snagged, she naturally attempts to work it free. If that does not prove

possible, she will eventually cut bait, that is to say, cut the line, or pull back hard enough to force it to break. If she is too stubborn to let go, she will waste valuable time and effort to no avail. Likewise, the most skilled and successful poker players know when to fold their hands and toss away their cards.

Well, you may not be into fishing or poker, but there is a valuable lesson to be gleaned from these examples:

Be smart (realistic) enough to assess the situation, and wise (flexible) enough to know when the plan is no longer tenable. As difficult as it can be to give up on something into which you have put a lot of time, effort and devotion, don't pour any more into a losing cause.

Then you need to get over it. Remember, the sooner you can free yourself from an impossible snag, the quicker you may be able to get started in a more positive direction.

Synopsis
Flexibility means meeting a challenge with persistence, finding a better way to get it done, or deciding not to do it.

Exploit Advantages
We're only human, and every one of us has weaknesses.

The best way to find a potential flaw is to look for it. And the most reliable way to profit from a flaw — or opportunity — is to anticipate and then exploit it.

For example, isn't it preferable to anticipate the failing underpinnings of a bridge rather than wait for it to fall? In other words, instead of driving over a potentially unreliable structure to see what happens, prepare an alternate route and have it ready when disaster strikes. This places you at an advantage over those who remain dependent on the bridge until the moment that it fails.

Since women are traditionally more nurturing than men, they may tend to overlook, forgive, or minimize the effects of other workers' poor performance. You cannot afford to fall into this trap! We're not discouraging you from being supportive and lending an occasional hand to deserving colleagues, only reminding you to be vigilant in recognizing vulnerabilities that may offer precious opportunities.

Anticipate
There's a reason why things go wrong.

Stuff happens and will continue to happen. The thing to keep in mind is that it happens for a reason. Imperfection is as natural to our world as hurricanes and earthquakes. For example, if your roof leaked last time it rained, it will probably leak again until the reason for the leak is discovered and repaired. If it isn't fixed, you can expect the roof to leak whenever it rains.

At work, just about everything that happens – good, bad and ugly – can be directly linked to people. When watching out for problems, what you see is what you can expect to get. Which means that people who usually do poor work will likely continue their pattern until something is done to correct it with assistance, training, supervision or replacement.

When a procedure, system or project begins to show flaws, it is because something (usually more than one thing) was not done correctly. System errors, missed deliverables and such signal a problem. The usual way to deal with such issues is to try to repair the snafu and negotiate for more time and resources.

If you're looking for an opportunity, seek no further! You can see that using more tape to support an item that is obviously too heavy for tape is, at best, a stopgap remedy. Finding a better solution demands a little research to determine the nature of the item: how large and heavy is it, where and how will it be used, and so on. If it keeps falling, something

more reliable than tape (maybe wire, rope, a table or special
container) is needed.

In other words, you can anticipate openings and opportunities from people who frequently cause problems, just
like when a warning flag is raised.

Recognize
Back in Chapter 4 we recommended viewing your weaknesses as lesser strengths to be improved. Now we'll turn
our attention to identifying imperfections, that is, faults
and errors that can be turned to your advantage. In the
workplace we're concerned with two kinds of weakness:
those we observe in behavior, and the ones that show up in
poorly conceived and executed plans.

Human weaknesses may be further divided into theirs
and yours. We have observed that women often go about this
process of observation and evaluation differently from men.

Say you're sharing a project with three other people and
something goes wrong. Until the cause of the mistake is
known, the odds are only one in four that it was your fault.
Meanwhile, all four of you are you are responsible for identifying and correcting it.

1. You: If it turns out that you messed up, acknowledge
   your mistake and learn from it. If the reason is unclear
   to you, seek clarification from your colleagues so that
   you will not repeat your error. Then move on energeti-
   cally, which is what men tend to do.

   Women, on the other hand, are more prone to person-
   alize their errors (even minor ones) and agonize over
   them. Again, no one gets anywhere without assuming responsibilities and taking risk. It follows that the
   more risks you take, the more mistakes you're bound to
   make. As long as these are interspersed with successful
   ventures, they are to be expected. So don't waste your
   breath crying over spilt latté.

2. Them: If someone else was responsible, be sure he understands what went wrong and how to make sure it doesn't happen again. Either way, the problem must be identified and corrected to allow the four of you to move forward.
3. Important: Whoever caused the error should acknowledge their mistake. Otherwise, they are unlikely to make the effort to avoid a repetition and may even try to blame someone else. Your response:

    Sample: "We're a team, so we'll work on resolving this together." But not: "Oh, don't worry, I've made worse mistakes."

    The point: Sure, everyone makes mistakes, but this one wasn't yours!

Plans

Poorly conceived and executed plans include the overwhelming majority of projects and deliverables. The larger and more complex endeavors are the most error-prone because of the number of details they contain, the people who contributed to them, and the conflicting agendas they reflect.

Remember to protect your own vulnerable areas (Chapter 3), because some of your colleagues and competitors are watching, probing, and waiting for you to reveal them.

Profit

There are several different ways to profit from an opportunity.

1. Offer to resolve a problem where you can gain visibility, recognition, and personal satisfaction.
    Caution: Be sure you know what you're doing in this case, because otherwise you'll look worse than if you had kept silent.
2. Help the responsible parties to fix it.
    Advantage: A smart and effective tactic is to be identified with a successful effort where you might not feel confident on your own.

3. Point it out to a decision-maker without getting involved.
   Beware: If the problem is too big and muddled for a
   quick fix, it may drag down everyone who works on it.

At this point you may want to ask,
   "Is it ethical to profit from someone else's misfortune?"
   There are several ways to address this question, depending on the circumstances and your point of view. Here are
a few questions that may help you to decide for yourself.
1. As already noted, mistakes are happening all around
   you. Is ignoring them going to help your career? What
   about the people who are making these mistakes – will
   your silence help or hurt them?
2. Should incompetent colleagues be allowed to continue
   sloppy work that negatively affects your company,
   group or you?
3. If you find that someone has made a mistake, might it
   be worth your while to offer to help them?

Synopsis
Anticipate, recognize, and take advantage of an opportunity before someone beats you to it.

Do the Right Thing
If it feels wrong, it probably is.
   Doing the right thing will help you more than it will hurt
another person. What is the right thing? Since no single
definition will apply to every situation, we'll offer a few
guidelines.
   With regard to your career, the right thing is what is right
for you. At the same time, consider the likelihood that whatever you do will displease someone. So you cannot filter all
of your decisions through everybody else's wishes. Otherwise you, and they, would for all intents and purposes be
paralyzed.

You might prefer to avoid stepping on the forest of toes that surround your every move. Since this may not be practical, however, you can content yourself by keeping the pain to a minimum.

Consider who your impending action is likely to inconvenience most, and evaluate the negatives: can you afford to annoy or alienate this person? How serious might that be? Could you discuss it with them in advance to seek a mutually satisfying compromise?

Rule number one, then, is to be aware of whom your action might hurt or anger, and decide if it is still worth doing.

Second, calculate what you stand to gain/lose if you do — or don't — pursue your plan.

Third, how much will you need to sacrifice in terms of time, effort, resources and risk in order to pursue your objective. Where would success or failure position you?

Finally, why do you want to do it, and how badly?

All of which bring us back to the original question: is it the right thing (for you)?

Example:

Your boss tells you that Daniel is falling behind on his documentation of a new software system, which could delay the scheduled rollout and implementation. He asks if you can pitch in to lend a hand.

You're willing to help, but have no idea of how much work may be needed or, for that matter, the quality of Daniel's work. So you ask for a couple of days to study the current state of affairs.

Scenario 1: Daniel's work appears on target, but he is clearly swamped with more than a single person can handle. So you offer to spend three weeks helping him catch up.

Scenario 2: Daniel's work is generally OK, but you determine a need to make some changes to the format in addition to assuming a portion of the work. Daniel welcomes

your participation, so you recommend an equal partner-
ship whereby you and he will share the responsibility.
Scenario 3: Daniel's work is clearly disorganized, suggesting that he needs both supervision and assistance. You
propose taking over the task, specifying that you will
need to reorganize and revise the work already done
in order to get it back on track. In this case, Daniel will
report to you.

   In all three of the above scenarios you are support
ive of Daniel and do your best to make him look good.
Needless to say, you will look good doing this.
Scenario 4: Daniel's is highly defensive about the project
and his work, and you suspect that he would resent
your involvement. So you avoid having any further contact with him.

Synopsis
If it smells and tastes good, it probably won't poison you.

Summary
Navigating your career through troubled times calls for a
steady hand and a clear head. With so many paths to choose
from, you must make wise decisions as to when to stop or
go, and in which direction.

   Sun Tzu has clearly and succinctly identified the conditions and parameters to be used in making daily and longer-term decisions, as well as the extremes to be avoided.
The keys are preparation, anticipation and flexibility.

   Remember, there are times when the most reasonable option is to step back.

Personal Profile
Jennifer Key — Balance humility with confidence.
Jen Key is an unassuming information technology devel-
oper with a strong professional drive and a charismatic na-

ture. Mother, wife and entrepreneur, she's one of those rare types who get along quite well on both sides of the Pacific.

I was born in Beijing and raised by my grandparents till I was about five years old. When the Cultural Revolution started, my grandparents had to relocate to a rural area, so I was sent back to live with my parents in Shenyang.

Mom and dad were scientists, and both were very busy with their scientific projects. I started elementary school at six, later moved back to Beijing, and once again returned to Shenyang to attend middle school. While in Beijing, traveling alone and from Shenyang by train was part of my life for summer and winter vacations.

My middle and high school years were filled with fun and friendship. Most of my classmates were from the same institute campus, where our parents worked for the Chinese Academy of Science. We spent lot of time outdoors playing games we designed after school. We were lucky enough to take the very first high school entrance exams following the Cultural Revolution. After high school graduation, I was blessed with the opportunity to attend college.

I majored in Electrical Engineering at the Dalian University of Technology, and was a stright-A student. I subsequently returned to Beijing and earned an MS at the Academy of Science. Although a bookish student, I began to realize that I wasn't interested in electrical engineering at all. Throughout graduate school I struggled painfully through several episodes of depression, recovering with the help of friends and family but in doubt of my future career path.

After graduation, I moved to Changchun to get married. There I worked for Jilin University, developing strong bonds with my students and falling in love with teaching. The opportunity to get to know and understand students was a wonderful experience.

At that time, traveling abroad was becoming a trend. Holding a belief that America would be the only place on earth to offer freedom for my soul, I worked really hard to apply to graduate schools in US. Two years of effort got me several admissions, but Louisiana State University was the only one to offer the financial support I needed.

My very first airplane flight landed me in San Francisco en route to New Orleans. Excited and lost, the world was stunning. But the English I thought I knew seemed a different language: not many people understood me when I tried to talk to them.

On campus, I experienced the unique pleasure of owning my own checkbook. But when the department secretary handed me my first paycheck, I didn't know what to do with it. Another new experience was the urge to earn and save money, counting each and every penny, because I realized that not having enough to buy food and pay the rent remained an ongoing risk. Life was tough, but exciting and hopeful, and believing that hard work would eventually pay off sustained and encouraged me to get through each day by learning more and more.

Cultural differences were reflected at virtually every occasion. You see, Chinese people are raised to be humble and less verbal, stressing team spirit over the individual. Whereas here in the US, individuals tend to make their voice heard. Even during classes students could interrupt instructors to raise questions whenever they wanted to!

Know yourself and your opponent

Sun Tzu's *The Art of War* was an early part of my education and working philosophy. One of the most important lessons he taught was about knowing yourself, your opponent and the environment in order to come out ahead. This, and the free-spirited attitude I discovered in the US, encouraged me to re-evaluate myself, my goals, and my options. "Who am I," I asked myself, "what do I want to do, and how can I position myself to get there?"

I concluded I wasn't cut out for hands-on work related to Electrical Engineering, since my passion and strength lay in working with people. And so, in order to utilize my engineering background and gear myself toward a more people oriented career, I switched my major to management information systems. The challenges of changing career majors were multiplied by the language barrier, which posed far more issues in business than in the sciences. But like they say, no pain, no gain, and the transition proved to be one of the best decisions I have ever made. As hard as it was, Sun Tzu's lessons helped me to realize that I didn't really know myself that well, to accept my own weaknesses, and eventually to see the positive side. Even having grown up with *The Art of War*, it took me until my thirties to apply it to my own life. And it helped.

Turn competitors into partners, pursue win-win situations:

After graduation, I worked for AT&T for five years. This experience provided me the experience of Corporate America, exciting teamwork, and a rich culture of diversity. It also convinced me that starting my own business needn't be just a dream; I could make it a reality. With a few business partners, we started O-Net International and then restructured to Ascentta, a fiber optics solutions company. Together we navigated the venture capital, join-venture, optical bubble, and other extraordinary periods.

I discovered that the business arena is a lot like a battlefield. Nevertheless, the best strategy is still a win-win situation. For example, some of our OEM partners are former competitors against whom we tried to compete with the same lines of products. After stubbing our toes on each other's feet, we eventually came to the conclusion that we'd be better off focusing our efforts on our strengths, and partnering on certain product lines. This enabled us to cover the entire spectrum with shared market segments.

Personally, I view win-win situations as a mindset, whereas the battle is mostly motivated by personal pride. When you stop to think about it, being generous to yourself and others is easier to say than practice. However, the effort is worth millions!

9. Managing and Leading

"One who does not use strategic planning and who underestimates the enemy will inevitably be captured." – Sun Tzu

The Master's advice is clear and unambiguous: In order to succeed you need a solid plan; and it is foolhardy to underestimate the competition.

With regard to management and leadership, military specialists have long recognized the differences between these important qualities. In civilian life, however, these distinctions are not always recognized. Let's take a look at the way they differ from one another.

Manage things

The heart of this difference lies in the contrast between material and human resources, that is to say, things and people. Managing people would only be practical if they were identical and completely known entities without individual personalities, experience, abilities and needs.

In brief, management is the way you allocate schedules and resources; leadership is how you influence your staff and team members to build loyalty, create cohesion and accomplish tasks. Many individuals and organizations assume that being a manager automatically bestows the grace of leadership, but this is simply not the case. Naturally, the two roles intersect and overlap at times, but effective managers and leaders need to recognize the differences. Business gurus Peter Drucker and Warren Bennis summed up the differences when they wrote, "Management is doing things right; leadership is doing the right things."

This is not to suggest that these skills are mutually exclusive, because many outstanding professionals excel at both. The main problem lies in the assumption that power equals privilege and ability. Maybe it's an ego thing that men, to a greater extent than women, take for granted.

A practical reminder of the difference between managing and leading might go something like this: the manager calculates the time and resources needed to achieve a certain goal, and the leader organizes and supervises the team to get it done. Of course, there's more to both than summarized here, but this is the core of how these two roles diverge.

You can see how a woman who grasps this distinction will have an inside track on colleagues who take leadership for granted. Why not you?

Fred is the Comptroller of a mid-western energy company. He hired Elaine as an Assistant Comptroller to demonstrate his open-mindedness, but was unable to release the reins. Whether through protectiveness or through distrust, he allowed several of the managers to second-guess Elaine's directives directly with him. As a result, it became common practice for the staff to question her decisions with Fred.

Elaine became frustrated and eventually resigned. Fred concluded that she couldn't handle the job or the pressure, and replaced her with a man.

What should Elaine have done? Perhaps if she had explained to Fred that she needed his support and for him to stand by (or to discuss with her) her decisions, she could have succeeded at her job. But she didn't, because she underestimated this need and felt that she could get by simply by working harder

Manage Things
The manager assigns tasks.

Managing focuses upon telling employees what to do and leaving them to get it done. Leading (next section) is the "getting it done" part where you teach, motivate and encourage the employees. Western corporations have been slow to recognize this difference, using corporate philosophies like MBO (management by objectives) and other sterile principles that attempt to force workers into predetermined slots. The fact that it doesn't work so well is at least partially evidenced by the success of Asian and other foreign management systems in contrast to our own. By way of example, we need look no further than the success of the Japanese automobile makers, in contrast to Detroit.

To be fair, some American companies have initiated changes, but as a nation we have a long way to go and are still in the process of playing catch-up.

Managing Work
Management skills are valued by most companies. They need managers to organize tasks, whether producing automobile parts, offering financial services, or whatever they

do to earn a living. Familiarity with virtually any kind of business may enable you to leverage your management skills to a wide assortment of related positions.

In recent years, the paths to managerial roles have been etched through marketing, product knowledge and technology. Such tendencies are based on the assumption that the ability to do something well equates to being able to control the process. However, this is no shoe-in, as many ex-managers can attest. One reason is an inability to step back and view the bigger picture; another is a lack of leadership skills.

Experience suggests that effective management requires training beyond product knowledge and sales performance. In other words, working your way up the ladder doesn't guarantee success at the next level. You may have earned the opportunity and feel that you deserve it, but you also have to be prepared for it to make it work. A successful manager needs to own a specialized set of skills. So if management is your objective, make sure to include both leadership and management as part of your preparation.

Why leadership? Isn't it possible to succeed as a manager without good leadership skills? In a few isolated cases, perhaps; but it's rare to sit behind the manager's desk without also assuming some responsibility for people.

When we come to managing the elements of your work, you need to weigh two main considerations:

Your authority and responsibilities as they (and you) are perceived by the people to whom you report and those you supervise. As a woman, you need to be assured of your responsibilities, authority and support.

Are you mindful and realistic about your strengths and lesser strengths, as well as those of your competitors?

Managing Expectations
Yours: Obsessive behavior is hard to take even in people we love. Ambition and success require balance and, yes,

moderation. We're not talking about excessive caution or rationalizations for not taking risks, only preparation and common sense. Jumping off a cliff may be a step too far, unless you're wearing a parachute or other reliable safety device to assure a safe landing. On the other hand, you owe it to yourself to allow your dreams and ambitions to soar as high as those of your male counterparts. Remember that reaching for the moon requires adequate propulsion, air, nourishment and, hopefully, a way back.

Your expectations also need to balance immediate and longer-term goals and objectives, possibly including family considerations.

Others': The expectations of your colleagues and employers are based upon what they believe they know about you: impressions, reputation, personal relationships and past performance. This is why you need to be aware of how you are perceived (your public image) and any differences between this and the way you see yourself. Hone your ability to read how they are reading you, and to emphasize or re-direct their impressions to suit your goals.

Communicating
It isn't possible to do the wrong thing right.

Everyone communicates, more or less, well or poorly. The ability to communicate effectively is essential to managers and just about anyone planning to pursue a rewarding professional career. Bear in mind that sharing and obtaining thoughts, information and instructions all require clear communications, and you're stuttering if you think otherwise.

As a manager, you need to be clearly understood by the people who work for you and those to whom you report. Misunderstanding in either direction can lead to disaster: as difficult as it is to do the right things well, it is impossible to do the wrong things right. So be sure that you have understood what is expected of you, and that your staff is crystal clear on your instructions to them.

If you haven't already noticed, the best communicators receive more than their fair share of attention and credit; poor communicators are often overlooked and, as a result, frustrated by the lack of recognition they attract. So don't blame the system if your career flounders due to questionable communications skills.

Weighing Factors

This seems almost too obvious to mention, and yet it is an area where many managers fall short.

One of the most common causes of a manager's success or failure is her ability to accurately assess the time and resources needed to accomplish an assignment. How many workers, what skills and training may be required, financial and other resources (such as systems access, proximity, hardware and software), authority to make and enforce decisions, and deliverable dates need to be assessed logically and realistically, not in la-la land.

Fact one: the majority of major projects fail, disappoint, and/ or are delivered over budget and later than intended.
Fact two: nearly all of these negative results are predictable.

Why do such things occur? Because wishful thinking and pressure are imposed over logical planning.

Most managers and executives are aware of project planning software like Merlin, Microsoft Project, Oracle Project Portfolio Management, Project.net, SAP RPM, and dozens of others. Many of these products offer ways to realistically evaluate how much time and resources will be needed to achieve a specified result. So what's the problem?

First, deliverables aren't always clearly defined; once work gets underway, new and modified details inevitably get tossed into the original plan. Which muddies the water and makes it difficult for the manager to accurately incorporate changes into the project planning software.

Next, input (dates, staff and so on) doesn't always account for subjective elements like individual work performance, deflections (reassignments of staff members), and judgment factors (if it usually takes one programmer 45 days to perform a task, how long will it take three programmers to complete the same job?). And small inaccuracies have a way of compounding themselves into major snafus.

Finally (for our purposes), women are often reluctant to confront or disagree with their bosses. Which means that a woman bent on proving her mettle may be susceptible to accepting an unrealistic schedule or poorly designed plan.

To protect yourself against unreasonable expectations and give yourself a chance to succeed, you must avoid this gaping crater. When asked to assume a certain responsibility, request all of the relevant details, chart it out on one of the software tools you feel comfortable with, and make your case. If someone disagrees with your projections, challenge them to justify their points. And if you are imposed upon to take responsibility for a plan with which you disagree, argue against it logically and unemotionally. At the very least, document your objections and disagreements for the record.

Example:
With regard to the current project plan, I estimate that the ZYX phase of the CBA deliverable will extend at least 6-8 weeks beyond the scheduled deliverable date. This is likely to impact overall project completion by 10-12 weeks.

Thinking Out of the Box
"...one who excels at the unexpected is as unlimited as Heaven..."
– Sun Tzu
Companies and institutions have their rules, taboos and cultures. Certainly you need to become familiar with these

to avoid obvious problems, which can range from annoying colleagues and higher-ups to grounds for dismissal. That being said, creative managers learn where extra pages can be added to the rulebook, or when it may be acceptable to think and act completely outside the book or box.

Common sense dictates a few basic guidelines when extending rules and tradition:

1. Wait until you have established a track record and some credibility before leaping too far outside the box.
2. Clear the air with your boss to avoid any unpleasant surprises.
3. Avoid negatively impacting your fellow employees.
4. Build support by including colleagues in your plans.
5. Document and be prepared to justify the advantages of your innovation.
6. Review how it may affect your career if accepted or rejected.

Don't allow yourself to be intimidated by conservative attitudes. Consider the book a starting point: ignore it at your peril; remain inside and you will be limited to a mundane existence.

Managing Time

"I wasted time, and now doth time waste me."

Shakespeare's provocative reflection reminds us that time can't be taken for granted and wasting it has consequences. Aside from health, time is your most precious asset. It's the one resource there's never enough of, and it can never be replenished or recycled. Women, who are called upon to nourish and care for family members to a far greater extent than their husbands, brothers or male colleagues, need to be particularly aware of the conflicts and stress threatened by multiple demands upon their time.

Chapters, even entire books, are devoted to providing tips on managing one's time. The central issue is that time management is closely linked to personality, style and priorities. For example, if you're overly sensitive about hurting people's feelings, you may be prone to spending too much time listening to their problems and cheating your own schedule. Likewise, failing to maintain and prioritize your to-do list is likely to create a frantic crunch to finish key deliverables on schedule.

To keep yourself on track and limit this kind of pressure, decide on what is most important to you, as well as which tasks to do in what order. Even if you prefer to keep your office door open to staff and colleagues, there are times when you need to shut it in order to address and complete important work.

Managing Health
Don't wait until you need a doctor.

In ancient times, rural Chinese doctors were paid a stipend (often in the form of food or clothing) by the members of their community to maintain their health. When someone fell ill, they stopped paying the doctor until s/he had successfully treated them back to health. In other words, it was to the doctor's interest to keep everyone healthy.

Today, in China and elsewhere throughout the world, doctors are paid to cure us when we're ill and fix our body parts when they malfunction. If you have children, chances are you closely supervise their diet and activities in order to protect them. But what about you — do you treat yourself as well as your family members? Are you equally concerned with your own needs and patterns, or do you wait until something's wrong before taking action? Is there someone else (a supportive spouse, sibling or close friend) on whom you can rely to share some of your family burdens?

Managing Assets and Deficits

No two people, male or female, share exactly the same strengths and challenges. In Chapter 6 we discussed how to maximize your strengths and lesser strengths as a woman while minimizing the downside of your weaknesses and liabilities. This is where you use them to succeed.

Synopsis

Manage your responsibilities, your relationships and yourself. Not people.

Lead People

The leader shows the way.

Sun Tzu encourages positive leadership skills throughout *The Art of War* through image, communication skills and interpersonal relations. These abilities can get you going in the right direction.

"The qualities of leadership remain, throughout the world of business, politics, and society at large, confused with those of management.

"To be a good leader you need to understand the difference between leadership and management. You can manage time, financial resources, production, inventory, conflict, and a slew of other items. If you try to manage people, you will fail." - A.D. Rosenberg, *101 Ways to Stand Out at Work* (Adams Media, 2009)

Leading

Effective leaders emerge from experience and training. Having a natural flair can be helpful, but you'll also need a flexible attitude, people skills, an open mind, and willingness to accept responsibility. These factors contribute to good leadership.

Why do many organizations confuse leadership and management functions? Probably for the same reason that we commonly find leadership listed as a managerial skill instead of a separate job title: it's because the distinctions are not clearly recognized. For this reason, a woman who understands the differences may enjoy a competitive advantage.

Let's compare the two concepts in a familiar household setting, and then expand them into the workplace:

Scenario 1: You tell your daughter, "You must get A's on your report card in order to get into one of the top colleges." Thereafter you monitor her report card, reminding her to improve when needed, and make sure she spends a certain amount of time each day on studying and homework.

Scenario 2: You tell your daughter, "You must get A's on your report card in order to get into one of the top colleges." Thereafter you discuss her successes and problems, and take the time to help her approach the topics she finds most difficult. You encourage discussion, propose approaches, offer guidance, encourage her efforts, and praise her successes.

Which of these seems more like leadership than management? The second one, of course!

At work, you are given a project to be completed in six months. Here is an abbreviated synopsis of the different roles you perform as manager and leader:

1. Management Scenario: You spec the deliverable from proof-of-concept to completion, including user testing and modifications. Then you decide how many of your staff members, and which ones, will be needed to complete the deliverable on time, request the necessary budget, and define assignments.

2. Leadership Scenario: You assign specific tasks to individual team members in writing and follow up with individual meetings. Next, you determine though ques-

tions and answers that they have clearly understood their assignments, and you address their concerns. After that you continue to monitor and encourage, providing explanations and training if needed. Overall, you put them in a position to succeed.

Leadership is about us, not me. A good leader makes optimum use of human effort. If you want to be a leader, think in terms of teaching, maximizing your team's skills, and moving toward a particular goal. The best leaders pilot by example, showing and explaining as they go.

To be a good leader, you need to recognize when to stand in front or blend into the group. There's nothing wrong with following another person's lead when appropriate. Leading doesn't always place you at the forefront: you can also be effective from the middle or the rear. Your responsibility is to recognize when someone needs your help and to provide it. You focus on identifying problems and their causes, and then move to resolve them. You cultivate cooperation and eliminate barriers. You provide both positive and negative reinforcement, clarify, allocate and energize – all qualities at which women tend to excel.

Roberta is a product manager at a cosmetics company. Her group is responsible for researching and creating new moisturizers. As of a year ago she hadn't come up with any fabulous products for several months, her budget was pared back, she lost two staff members, and morale in her department had reached an all-tine low.

Instead of complaining or getting down on her staff, Roberta took on the reigns of leadership in an energetic manner. She divided her employees into two teams, provided slightly different guidelines to each, and helped them to get started on their research projects. Over the following weeks she divided her time more or less equally between them, offering suggestions and encouragement.

When one group came up with an interesting idea (a moisturizer that could also function as a sun screen), Roberta reassigned the second team to play the role of upper management, asking tough questions and challenging unsupported assumptions. Then she brought them back together for a discussion, divided them along different lines into production and sales teams, and helped them to prepare detailed R&D (research and design) and projected marketing presentations.

As the preparation neared completion, Roberta quietly moved further to the rear of both teams, encouraging and providing guidance with a softer voice.

As a result, Roberta's employees gained confidence in themselves and one another, bought into the product they created, and elevated her department into one of the most productive in the company.

In reality and practice, many managers and executives assume that authority easily transfers into leadership. However, the assumption that experienced managers are automatically qualified to lead is off the mark. Equally wrong is the notion that people can be successfully managed instead of led. These misconceptions open the door to women who understand that leadership and management roles are different and require separate skills.

Leadership requires a positive (self and public) image, good communications skills, and experience. Good leaders recognize that leading is about teaching, showing, encouraging, inspiring, and doing. These are commonly classified as "people skills," where women often enjoy an advantage over men.

Male images dominate the overwhelming majority of leadership roles. You only have to turn to a TV news channel or open a newspaper to confirm this tendency. It's not so much that men are more skilled or talented leaders than women, although they do create a splendid show of confidence.

How can women overcome the historic image of large and confident-looking males riding mighty steeds or motorcycles? One way might be to saddle up an even larger horse or motorcycle, although a better alternative would be to become a darned good leader, and to prove it with measurable results (refer to Chapter 8).

In our view, women are more natural leaders than men: they're better teachers because they know how to listen; more nurturing because this has been their traditional role; more practical and down-to-earth; and more experienced in providing guidance and serving as role models to their children.

Credibility

Leadership skills need to be balanced with credibility, a condition earned through building a reliable track record. Female employees face the same issue as their male counterparts, that of gaining hands-on experience. However, getting started is usually a greater challenge for a woman. The reason is twofold:

1. Personal attitude: Few women envision themselves as leaders. If your career is lagging due to uncertainty about your leadership skills, self-confidence, or reluctance to take risks, review your career objectives to determine if a turnaround is needed.

2. Public attitude: Until more women become recognized as competent and reliable leaders, assuming roles of leadership will remain an uphill battle. By all means, draw inspiration from Hillary Clinton, Margaret Thatcher, Angela Merkel and hosts of others, but remember that the majority of effective female leaders are neither rich nor famous. They're just competent, devoted, hardworking people, pretty much like you.

Leaders aren't born – they're educated
The most direct path to leadership is, of course, through
leading. Take it in small steps at first to build your confidence and grow accustomed to the role. Co-lead with someone – male or female – who is more experienced, and gradually assume additional responsibility. Try to avoid infighting (you don't want to be identified as a power-player), and
negotiate roles with your partner. Volunteer to do certain
things, and let the other guy or gal receive most of the glory,
even if they're not inclined to share. The important thing is
to be associated with leading a successful venture. That's
how you launch a track record and build credibility.

If you're ready to take on a leadership role, set the odds
in your favor by assuming responsibility for a modest project without too many loose ends. Analyze the objectives,
resources, schedules and details to be sure that the goals
and deliverables are realistic (no one has yet run a marathon in under two hours). Assess your ability to pull it off,
and any additional resources (staff, equipment, funding,
etc.) you may need. No matter how strong your confidence,
solicit the opinion and advice of an experienced individual
or two. A degree of risk lies within every responsibility you
assume, but smart players rarely bet against the odds.

Address each member of your team individually. Be sensitive to any attitudes from males or females that may be
related to your gender. A common theme throughout this
book is to remain aware of potential barriers without either
assuming them on the one hand or becoming apologetic
on the other. Ignoring situations because they are unpleasant reduces your ability to deal with them effectively when
you must.

Overall, treating colleagues and collaborators fairly,
lending a helping hand when needed, and being firm and
supportive when appropriate, will win over a majority of

reasonable people and a few skeptics to boot. As for open-mindedness, the best anyone can do is to offer folks a reason to unlock their thoughts. The rest is up to them.

Qualities

There are three widely recognized paths to leadership:

1. Possessing attributes: A pleasant, outgoing personality, self-confidence, ability to observe and communicate, and willingness to take charge are helpful resources.
2. Rising to the occasion: When opportunities arise, strike while the iron is hot. Otherwise someone else will, and all your wonderful qualities may go to waste.
3. Learning: The ability to learn is the other side of the leadership coin. Only by recognizing what works (and what doesn't) can you pass along the message to others.

Leadership Principles

Here's our list, and we encourage you to come up with others:

1. know yourself;
2. recognize and improve your lesser strengths;
3. get to know the individual members of your team;
4. enable them with guidance, encouragement and the right tools;
5. communicate clearly;
6. be courteous and responsive;
7. take responsibility, especially when things get tough;
8. share credit;
9. act in a timely manner;
10. be a role model.

Bear in mind that the attitudes of your employees and colleagues reflect their confidence in you. Credibility will influence their willingness to give you the benefit of any doubt.

Synopsis
Lead people by example, and position them to succeed.

Solve Problems

Problem solvers are worth their weight in pearls.

In separate or combined roles, managers and leaders share the need to recognize, define, and resolve the large and small problems that accompany responsibility like fish in water. Here are our recommendations.

1. Develop an honest understanding of your assets and potential deficits, especially with regard to the current assignment. Do you have the necessary skills, experience and resources to get the job done?
2. Maintain composure and consistency. Instill a harmony of purpose and action with a reliable pattern of behavior and expectations.
3. Show and share enthusiasm with everyone around you. Those who respond are your potential allies and assets.
4. Commit yourself to excellence over the short and long haul. Continue to learn and build toward your career objectives.
5. Learn to reach without overreaching. Focus on your priorities and don't be lured into probable dead ends. Stay true to yourself.

Bearing these suggestions in mind, the best way to deal with problems is to avoid them in the first place. While no one is entirely bulletproof, some people have a knack for circumventing sticky situations. Watch these individuals and learn from them. You'll find that luck, for the most part, is the direct result of keeping your eyes and ears open, and being realistic with yourself and your environment.

The other side of avoiding problems is making sure that your successful efforts are known and appreciated.

Why? Because the planning, time and effort you devote to anticipating and circumventing potential issues are far less obvious than the ones that land on your doorstep and need to be resolved. The simplest way to deal with this subtlety is to identify your vigilance in a status report, e.g.:

I noticed that the time allotted to Step P-17 (3 days) failed to take account of the need for ABC as a prerequisite, for which reason I had the ABC phase addressed and completed in advance. This action avoided a time-consuming bottleneck that would have delayed at least six technical staff members.

Synopsis
Most problems are opportunities in waiting—if you're ready.

Establish Priorities
What's worth doing first, if at all?

Priorities are essential to managers and leaders in allocating time and resources. Without priorities you can put in long hours and honest effort to little avail.

Start by evaluating immediate and then longer-term goals. Then decide which of them is worth your time and effort, and in what order. Consider the consequences of doing or not doing a particular task, and how urgent it may be. Are some tasks dependent upon others?

You may find the following criteria helpful:
1. Is it a requirement?
2. Does it contribute to an important goal?
3. Does it have an impending deadline?
4. How much effort will it require?
5. What is likely to happen if I do/don't do it right away, later, or not at all?

Obviously, if the answers to 1) and 2) are negative, you can probably set it aside. Item 4) might depend on 3), rendering it too difficult to complete within the available time frame. Then again, the last item on the list might cause you to reconsider the first four.

The two main categories for priorities are by importance (something that must be done) and urgency (it must be done immediately).

Example 1: importance
Group leaders for the main tasks that make up a forthcoming project must be chosen before work can begin.

Example 2: urgency
The project is currently in the planning stage; at present, the architecture and funding are the most pressing needs.

One final thought on the topic is that if you fail to set priorities, someone else may do it for you. Then you're stuck with opinions, values and interpretations you may find hard to swallow.

Synopsis
Determine where you want to go and how to get there from where you are. If you can't get there from here, then you need to be somewhere else.

Meet Needs

Everyone has needs
Sun Tzu's armies required adequate food, clothing, transportation, equipment and weapons. As in Sun Tzu's time, civilian needs are more subjective, as most of us are expected to supply our own clothing, food, and transportation between home and work. Jumping forward to our decade, work tools like computers, writing equipment, desks,

chairs and telephones are normally provided by our employers. So what do we need?

Ask

In the modern workplace most needs are job-related: adequate resources (staff, funding and time), access to information, expert assistance, and managerial support are among the most common examples. The need for help, which we all experience on different levels, is the one that we are most reluctant to express or request. Why? Because we don't want to appear incapable of meeting our responsibilities.

Males and females typically react differently to needs. Men may not care to to admit it openly, but they may still seek out someone with more experience and ask what they would do in a similar situation:

"Just to potentially expand my options," or

"I have it under control, but I'm always open to a more efficient way."

Many women, on the other hand, lean toward feelings of inadequacy and blame themselves for not being able to accomplish difficult tasks on their own. They forget that there are times when anyone could use a hand or words of wisdom. Under stress, asking for help may seem like a sign of weakness, as opposed to a logical way to deal with a question or problem.

The best way to meet your needs is to express them to your boss, or perhaps a colleague whom you respect and trust. Even if they aren't knowledgeable on the topic, they may be able to refer you to another resource.

Obviously, this is not a path you want to wear into a well-marked trail, but from time to time, it makes sense to ask.

Answer

Sooner or later, everyone around you shares the need for help. This is where you may be able to assist another person's need, a great way to build credibility and respect. Even if you are not an expert on the issue, sometimes all the other guy or gal needs is a friendly ear. And, by and large, women are better listeners than men.

In other words, just by being receptive you may be able to offer valuable assistance to a colleague in need, especially if they are worried or upset about their situation. You might even guide them toward resolving their problem by viewing it in a different light, or referring them to someone else.

Synopsis
Don't forget about your needs, and don't be afraid or embarrassed to ask for help.

Summary
Now that you're familiar with the differences between managing and leading, you're ready to apply these skills to your career plan.

Remember to use your management strategies for things while becoming a leader of people. It all boils down to doing the right things right. Since women tend to excel in understanding and motivating people, you already have a leg up on your male colleagues.

10. Moving Forward

"Know the enemy and yourself, and your victory will never be endangered; know the weather and the ground, and your victory will be complete." – Sun Tzu

The current chapter focuses upon geography, or shape. Sun Tzu was referring to the conditions that confront an army, and the commander's decisions about when, where, and how to advance. Modern world adaptations apply to the circumstances and conditions that lie before you as you plan your moves. As usual, we confront the fact that women are more susceptible to criticism within male-dominated environments and must exercise caution before committing themselves to major steps.

Our teacher defines six types of terrain and an equal number of causes of failure. To these he adds good discipline, and knowledge of yourself and your opponent in assessing when to advance or temporarily cool your heels.

Evaluate Conditions

Sun Tzu's definitions of terrain include:

1. Accessible: A friendly playing field that offers equal opportunity invites quick action. If your company's P&L is favorable, your department within budget and the boss in a good mood, make your move before one of your colleagues gets there first. This is where hesitation may lose a potential opportunity.

   Likewise, launching a good (if imperfect) product or service into a void of competition may establish a foothold before the other guys arrive (as long as it is not a total bust).

2. Entangled: Some situations are a cakewalk to enter but quicksand to leave. So if the door appears wide open, be sure there isn't a nasty trap awaiting you on the other side.

   In warfare, it is easy to march downhill into the valley, but hell to get caught there if the enemy holds the surrounding hills. Within your company, a promotion may seem inviting, but check to see if the responsibility it entails isn't being avoided by your colleagues for a darned good reason.

3. Temporizing: No-one's land is where you'd rather draw your competition than your own resources.

   Don't get conned into a no-win or dead-end job under the guise of "It needs a woman's touch" (what

doesn't?) or some such nonsense. Women are all too
willing to accept difficult tasks in order to prove themselves, but make sure it's doable, not a deep pit from
which escape is nigh impossible.

4. Constricted: Situations with little room to maneuver and
   difficult to navigate must be scrutinized before entering.

   If you replace someone who has already established
   tight working parameters, budgets and deliverables,
   you may have to live with them. In which case, you'll
   need to know if and how they succeeded.

5. Vertical: If the high ground is available, grab and hold
   it. If it is already occupied, look elsewhere for better opportunities.

   The worst position you can occupy is from the bottom looking up at an ensconced competitor. This, more
   often than not, may be the wrong fight.

6. Remote: Distant or vague objectives and conditions are
   difficult to assess. Try to obtain an overview of the terrain, or at least a thorough scouting report before em-
   barking.

   Hindsight and logic remind us that Germany's long-
   distance invasion of Russia was doomed to failure.
   Likewise, seeking advantage over a competitor is mean-
   ingless without an image of their products, services,
   practices, reputation and customer base.

Synopsis
Before deciding which shoes to wear, find out if you are going dancing, skiing, or walking in the rain.

Create Favorable Circumstances
Control freaks burn out; blind optimists get burned.

The US has the power to decimate Iraq, Afghanistan, and
dozens of other nations that trouble us. But since such total
destruction is not a viable political option, our involvement

in these far-off and poorly understood lands is plagued by conditions that evade our control. As time passes, our losses mount and progress lags. When you think in terms of win and lose, there is little to recommend our ongoing involvement. Yet given the nature of our commitment and sacrifices, we seem stuck in an unpopular battle with no end in sight.

To paraphrase Sun Tzu, control the things you can and closely watch the things you can't. If formidable barriers block your way in one direction, seek to advance in another.

For us this means navigating carefully. If you think the job will need more staff, budget and time, be willing to argue for them. If the powers that be refuse to give you what you need, state in writing that you don't believe the job can be done successfully within existing parameters. In other words, stand up and state your case even at the risk of getting someone annoyed at you. This is a better option than agreeing now and failing to deliver later.

Women are notoriously reluctant to confront or strongly disagree with their employers. That's why some managers will assign the more questionable tasks to women; they know that a man is more likely to reject an obviously bad deal. Well, you don't have to play the role of fall-gal to anyone's unrealistic agenda. That would be like agreeing to drive the train without brakes or alternate tracks, completely out of your control.

Remember to depersonalize your argument:
"I don't think the job can be done without additional..."
is preferable to
"I can't..." or "How do you expect me to..."

Synopsis
Exercise control over the things you can, remain wary of the rest, and do your best to balance both.

Take The Right Path

Let us stay aboard our train for a while longer to apply Sun Tzu's advice. Like you (and anyone else), the train needs certain basic attributes to continue rolling safely to its scheduled destination.

These principles seem so simple when you break them down into essential elements:

1. power: You must have the minimum resources to propel your movement forward and sustain momentum;
2. solid ground: The company's infrastructure must provide the means to provide staff, acquire and disperse financial resources, and so on;
3. clear path: Obstacles, such as inadequate staffing and financial resources, internal opposition, and restricted authority must be cleared from your path before you can make progress.

The right path should also include your career interests: will the assignment help you to advance toward your objectives (within and outside the company), or is it a dead end? Do you find it interesting and potentially rewarding? If not, perhaps you should look for an alternative track.

Example:

The boss: "I need you to set up a satellite office in Houston. Here's a list of what you should need, including schedule and resources."

You: "Great! I'll verify your time and resource estimates and get back to you by Thursday morning."

Instead of accepting someone else's life-support system, you're going to make sure that their expectations are all reasonably supported. Note that you have labeled their numbers as estimates subject to your confirmation.

Synopsis
Whatever path you accept or choose, make it yours.

Avoid The Wrong Move

Sun Tzu identifies six criteria for avoiding bad decisions:

1. Opposing odds: If you are a relatively recent hire, don't go up against an established and respected member of the department. Likewise, resources that will last no more than a couple of months cannot reasonably be stretched into a year. If you're being pressured into accepting a task despite facing formidable odds, it is reasonable to wonder why you, a woman, are being singled out for an honor your male colleagues won't touch.

   And if you're tempted despite these warning signs, question your own motives: are you determined to show everyone that you can beat the odds, pull off the impossible? Because even if you do, what will they give you next?

2. Poor management: If your boss is disorganized, unreasonable or unsupportive, examine any special requests or unusual assignments as if your job depended on it.

   In small companies, owners commonly insert sons and daughters into positions of authority for which they may have no interest, aptitude or background. In larger organizations, older or dead-ended managers may be resentful of youthful employees, or simply worn out and waiting for the calendar to announce their retirement. If you work for someone who does not support you and your assignments, focus on escaping to a more supportive environment.

3. Inadequate staff: The best management in the world can't function without good people. Of course, great managers and leaders are unlikely to get stuck with duds in key positions, but with frequent personnel

changes and budget crunches, you cannot afford to take this for granted.

So if you're asked to lead a team, find out as much as possible about their track records, including previous experience, attitudes, and ability to work cooperatively. Even as a leader you will need support from your staff to succeed at whatever you've been asked to do.

4. Conflicting agendas: Especially in larger organizations, upper management and executives are known to pursue their private agendas individually or in little cliques. This can serve to undermine the corporate policy makers as mixed signals filter down to middle management and professional staff.

   From your perspective, watch to see if policies and practices, as they are communicated to you, appear disorganized, contradictory or inconsistent. If so, this is a company you may want to leave as soon as you can find a better place to spend your working hours.

5. Low morale: Mismanagement, conflicting agendas and poor leadership will make it difficult for managers to cooperate and trust one-another. Obviously, this is a poisonous environment to all concerned.

6. Incompetence at the top: When the leader of an organization, division, group or project makes major tactical errors, the ship has sprung a leak. In war this equates to attacking against superior forces in fortified positions. In business, it would be like introducing a new cola to compete with Coke and Pepsi without multi-million dollar product research, production, design and advertising.

   An example that comes to mind was the attempt by a major California wine producer to penetrate the French consumer market. They advertised like crazy but paid no attention to the question of taste. Another was a US automaker's promotion of their Nova model in Mexico and other parts of Latin America. Apparently, their

Spanish-speaking employees didn't tell them that not too many Latin men would want to drive a car whose name translates in their language as "No Go."

Synopsis
You cannot avoid all possible mistakes, but you can cut down on the most obvious blunders.

Be Disciplined and Patient

Discipline
The best plan will fail if not pursued with integrity and discipline.

Good discipline and fair treatment net high rewards for managers and leaders. Be sure to reward workers for what they do right and correct their errors. This provides the confidence to move forward and the motivation to offer their best efforts.

Give credit where it is due, not to yourself. Shine the spotlight on your staff and colleagues, and they will follow you enthusiastically. Combined with competent leadership, clear communications and solid management decisions, this is where success can be elevated to its highest probability.

Patience
"Life is a marathon, not a sprint." – Dr. Donald A. Redelmeier
There are favorable and unfavorable times for action, some more evident than others. Given the impossibility of addressing every possible situation, we offer a few obvious examples:
1. Not everyone can move at your speed. Consider as well the quality of their work.
2. Friday afternoons are usually not the best time to ask your boss for an important or difficult decision, especially she happens to be leaving for a long weekend.

3. If a competitor gets there first, regroup to consider if
   your intended move is still in play, or if you'd be better
   off changing direction.
4. Do you have all the ammunition you need, or might a
   little more preparation reinforce your case?
5. Never give up on yourself when things don't move as
   quickly or smoothly as you hoped. If you aren't patient
   with yourself, how will you be with those around you?

Synopsis
Discipline and patience are the yin and yang of efficiency.

Summary
Military and workplace decisions drill up to knowing how
to evaluate the conditions in which you are operating. A key
element is being able to influence important circumstances
by observing, analyzing, and applying common sense. Another is to stay clear of obvious and damaging mistakes.

Stay disciplined: some calculated risks might be necessary, but leave the bonehead moves to boneheads.

11. Deployment

*"Seize something the enemy cherishes and he
will conform to your desires." – Sun Tzu*

Chapter 11 translates into something like "Nine Varieties of
Terrain." We shall apply this definition to:
1. conditions and circumstances to consider before committing yourself (time, resources and reputation) to an
   endeavor, and
2. how to maximize your likelihood of success.

If these themes sound familiar, stay with us. Many Asian
historians consider this to be *The Art of War*'s most discon-

nected chapter in that it introduces certain topics incompletely or out of context. We hope that by integrating a few of Sun Tzu's thoughts from other chapters we'll make this chapter as helpful as possible. Our point is that a woman who is aware of the environment she is about to enter will be better able to match her strengths against the opposition's vulnerabilities.

Pick Your Turf

We begin with an overview of the nine fields of play described by Sun Tzu, and how to profit from (or avoid) them.

Home turf

In ancient China, civil wars between neighboring kingdoms were common. Nor have civil wars fallen out of flavor in our relatively new millennium. At work, perhaps you're one of several people seeking the same assignment or promotion.

Overall, keep your personal ambitions to yourself as much as possible. Reveal only as much of your intentions as you must to those who need to know: your boss, a job interviewer and so on. Show enthusiasm for your intended role or position, and balance humility (a stereotypic female tendency) with confidence (a common male excess).

"But wait," you may protest. "I'm an individual, not a statistic; why should I compare myself to women who have nothing to do with me?"

Good question! In fact, no woman (or man) we know conforms precisely to gender stereotypes. Yet as long as they exist (and make no mistake, they do) they will influence people's attitudes toward you when they first meet you. Maybe this isn't how things should be in the world we live, work and play in, but ignoring reality is a mistake. Changing unfairness and injustice requires a tight handle on the way they actually function. Face it, there is no glory to hiding your head in principles while leaving your posterior exposed.

When you get the job you want or temporarily settle for second best, perform your duties conscientiously and thoroughly. Continue balancing modesty (giving credit to those who merit it) with confidence (can do!). Even if your boss is woman, don't necessarily expect to be recognized as an exceptional contributor. Be mindful, however, that women managers and executives may be watching you more closely than you realize.

If the opportunity to network (Chapter 13) arises, grab it! Anytime a higher-up seems willing to mentor you, be worthy of their interest, and remain loyal to everyone who helps you along the way.

As for the internecine battles that take place around you, pay attention to them without (if possible) taking sides. If you're pressured by an individual or group, ask the advice of a senior staff member or manager before choosing.

Neighboring turf

When you move your feet into another side of town, step lively. Don't be lulled into pussyfooting about because you aren't far from home, or you may stub your toe.

In a new job, move carefully but quickly to own your responsibilities. Until you have established yourself, you're still on unfamiliar ground.

If you are in competition with another branch or service, give potential clients a reason why it could be to their advantage to consider dealing with you. Establish toeholds, and then work to expand them until you reach equal footing. Then move to dominate.

Don't read this as "caution to the winds," but as timely and persistent movement. Hesitation on this ground will likely send you spinning backward into never-land. Yes, preparation is essential, but half a loaf and a kilometer are better than a full loaf but only a few meters. Your future may not ultimately depend upon this segment of your journey, but your performance here will always be a part of your track record.

Why do we emphasize swiftness and persistence at this stage? Because (back to stereotypes for a moment) females tend to be more cautious and less aggressive than most males. Remember that folks are watching and making comparisons. Once you prove yourself worthy of the task, your reputation will begin to rise above some of these preconceived attitudes.

Competitive turf

Competition can be a fair and healthy process, or it can be extremely ruthless and cutthroat. Our advice is to prepare for your campaign by assessing the competition. If another candidate clearly has the inside track and is backed by more experience and preparation, reconsider whether this attempt is worth your effort.

Alternative 1: Approach that individual, acknowledge their credentials, and ask them to advise you how to prep for a similar opportunity.

Alternative 2: Ask them to consider selecting you for their team to enable you to learn from them while making a valuable contribution.

On the other hand, if you appear at least equally qualified, try to match your strengths against their weaknesses, and come up with a rationale that acknowledges your minuses in a favorable light.

Example: "I'm in the process of getting up to speed on the math used to calculate bonuses and commissions."

Competition between candidates for a new position, promotion or choice assignment is often subjective. Assuming that all concerned have valid credentials, other factors (like seniority, popularity and politics) may enter.

Sun Tzu's advice to fight only the right fight prevails throughout his strategies and tactics. In the current instance, it means competing actively when you have a reasonable chance of winning, rather than going against the odds. In general we agree, but there are exceptions. If, for instance, your desire to compete for a project is so intense that you are willing to risk losing (or have little to lose by not winning), by all means go for it! Life isn't always about playing it safe and waiting for a chance that may not come. At least you won't have to wonder afterward what might have happened if you had given it your best shot. Just be mindful of the likely scenarios in either case.

Either case? That's right, because if you do come out on top of a hotly contested effort, people (especially the ones you beat out) will hold you under a microscope that may exaggerate your mistakes while minimizing positive results. So if you decide to buck the odds, do it with an understanding of the potential pitfalls alongside the benefits, whatever the results. Be ready to justify (mostly to yourself) that you did indeed make the right choice.

Alternative: Depending on what suits your personal style, you may decide to vie for a less competitive assignment, one where you will enjoy the relative freedom to build yourself into a rising star. But first, find out why the tougher competitors didn't want it.

Neutral turf
We interpret this category as the open market, to which everybody has equal opportunity. Whether thinking about individual ambitions or a company strategy, market domination is least likely to be attainable and virtually everyone has access to a piece of the pie. Since there isn't much here to favor women, we'll move along to more fertile ground.

Overlapping turf

This is what Sun Tzu calls ground that is contiguous to several states. His concept is that being able to control this area will win the support of nearby neighbors.

The application to our times can be surprisingly useful. Suppose your company had recently taken over a small but vital firm and was exploring ways to integrate it seamlessly into the existing setup. In our scenario, you're the head of a department or division seeking to meld the acquisition into your group. Add the complicating issue that two or three other department heads are also making pitches to grab hold of the new prize; and they are led by men who consider themselves entitled. Is it in your interest to go head-to-head against the guys, or might there be a better way to pursue your objective?

What if you were to approach a group leader — preferably one whose abilities and staff complement yours — to team up with a joint proposal? All other things being equal, wouldn't this approach create an advantage over the other squabbling competitors? Moving right along, one of the other guys might begin to realize he had been outflanked and ask to join in, strengthening your position even more.

Suppose they all decided to participate in your joint venture, how likely would that be to win? Always keeping in mind, of course, that you, as the original organizer and point person, retain the prime seat in the coalition.

From your conciliatory, non-provocative approach, you've created a win-win conclusion over antagonistic infighting and potential acrimony. The perfect woman's touch!

Committed turf

There comes a point where commitment can no longer be avoided if you intend to actively pursue your dreams. Having worked your way into the land of responsibility and credibility, you target larger trophies and are prepared to make a major effort to obtain them. No longer sitting in your comfort zone, you're out there with the big girls and boys, a force to be reckoned with; but then, so are they.

What's different from the competitive turf we discussed a bit earlier? Well, the games are played pretty much the same by matching strengths against weaknesses, but here the stakes are higher and more costly. Winning may require alliances, compromises and debts; losing could cast your career into a negative light. Thus you must advance with careful consideration, because having penetrated this far, there may be no retreat. Win or lose, you'll be subjected to more scrutiny than ever — it goes with the turf.

Rugged turf
In Sun Tzu's time, military forces had to contend with mountains, rivers, forests, swamps, and who knew what else? The metaphor extends to known and unpredictable issues of a financial, legal, political, cultural and technical nature. Nor should we overlook conflicting human attitudes, personalities and agendas.

As difficult as it can be to navigate the predictable bumps and avalanches along the road, it's pretty much impossible to anticipate all of them. Which is why your game plan needs contingencies to cope with the unexpected. On the one hand, you must prepare for sudden roadblocks and hazards; on the other, you cannot afford to let their specter stop you from moving forward.

When Lewis and Clark crossed America in the early 19th century they faced and overcame completely new and unexpected obstacles and dangers. Their ability to plan, anticipate and improvise enabled them to travel from the Atlantic to the Pacific and back again.

Since then we've visited the moon and sent space vehicles to the far reaches of our solar system. And yet, some forty percent of Americans have never lived outside their places of birth. Fair to say that progress in a world dominated by a hard-core population adverse to change rests on the shoulders of those who are willing to take risks.

What might this mean to women? One quick observation suggests that lots of opportunities are lying about and waiting for someone to pick them up. Well, true enough, but of course there are other factors cluttering the way. For one thing, aggressive types often find themselves locking horns with other ambitious men and women in the largest cities and corporations. That's where most of the action is to be found, and why we call it rugged turf. So if you're looking for a smooth ride, you may be on the wrong track.

Men are accustomed to playing rougher games than the majority of women. They are oriented toward competition through sports and social contact. Women's sports are definitely coming into their own but have yet to achieve equal billing. Another factor to consider is that, to date, most sports are segregated by gender: girls compete against other girls, not boys. So when you take the gender bounce into so-called equal opportunity, remember that the tools and equipment you grew up with may be different from the ones used by men.

Which is to say, the game has changed, and the transition demands a little more of you.

Entrapped turf
Professional women surrounded by men understand the concept of being surrounded, as do members of racial minorities who are scantily represented within corporate environments. Sun Tzu's definition of entrapment relates to situations where a strong commitment has been made, retreat appears a poor option, and you are outnumbered.

What to do? Above all, don't panic, because you will thus abandon any possibility of success and look bad in the process. Your viable options boil down to:

Change your course onto a new track. For example, forge
an alliance to strengthen your position, or alter your objective in a way that seems more viable. You might even down-
size your plans into a more manageable and tenable path.

Seek to alter the conditions that threaten your objective,
such as having resources and support diverted from another venture into yours. Naturally you would have to justify
your request, but no one said it would be easy.

Subvert or co-opt the opposition. This tactic is similar to
the one used on overlapping turf, except that now you are
negotiating from a position of weakness.

Look for a graceful way out: no use tossing good money
after the bad. Backing down by admitting that you made
an error of judgment, or that conditions have impacted the
situation, may be preferable to sinking with the ship.

Dire turf

The soldier's credo is that if you're going down, you might
as well take as many of the enemy with you as possible.
Fortunately, we don't need to construct quite so desperate
a scenario for our readers. However, there are times when
an extreme effort is the only remaining chance of survival.

In today's China, young college graduates, many with
advanced degrees, are willing to put in sixty or seventy
hour weeks in order to establish themselves in a new company. In our experience, few Americans (and far fewer Eu-
ropeans) are willing to sacrifice their time and effort in the
same manner. Instead, we demand benefits and recognition
as if entitled to rewards and success, overlooking the dedi-
cation needed to come out ahead in an increasingly competitive marketplace.

Women, on the other hand, are more prone to pay the
price, often to the point of being exploited by less than scru-
pulous employers. Indeed, a fine line exists between exploitation and reasonable sacrifice.

Our message to women is that within an environment of entitlement and expected privilege, a willingness to go the extra mile (not marathon) can position you as a reliable resource to those who appreciate your efforts and results.

Synopsis
Understand the environment in which you function. Preparation and familiarity make any ground friendlier.

Confuse Opponents
"Keep your friends close and your enemies closer." – *Sun Tzu*

Throughout *The Art of War* Sun Tzu promotes the tactic of keeping your opponent off-balance. This will help you to avoid their strengths and enable you to penetrate their weakness. In our world, it applies primarily to competition from other companies, as you wouldn't want to overtly deceive your colleagues. But even your co-workers don't need to know your ambitions and plans unless they're helping you to achieve them. The rules of thumb are:

1. Disrupt the opposition's timing by feeding them misleading information.
   Example: Leak notice of a different software update than the one about to be released by your company.
2. Confuse their strategies by doing what they least expect.
   Example 1: Offer to collaborate with your main adversary. This will take them off balance and possibly reduce any animosity they may have toward you. What's in it for you? The likelihood of reduced tension, insights into your competitor, and a win-win situation: your colleagues and boss will respect you for your openness and self-confidence.
   Example 2: Praise the competition in a manner that works to your advantage: "XYZ has been the standard

for years, which is why we had to offer additional options while matching their quality."
3. Create confusion by representing your strengths as
weaknesses and vice-versa.
Example: Improve a product flaw to make it stronger
than the competition's former advantage.

Synopsis
Information is power; the less your competition knows
about you, the better you can time your moves.

Energize

She who hesitates beyond preparation and assessment will
make no progress. When you stop to think about it, it's
easier to conjure reasons for waiting than moving forward.
Gear up, do your homework and get it done.

If you're working alone, give yourself pep talks to keep
the positive energy flowing. The further you advance, the
greater the danger that uncertainty and anxiety may con-
spire to slow or divert your progress. Try to discuss your
work, concerns and progress with a colleague. If it's someone who is also working on their own, you may be able to
provide a sense of teamwork to one another.

As a manager or team leader, work to keep your staff's
spirits up by listening, encouraging and leading. Be the
kind of example you would want to follow, and they will.

Create a hierarchy of teamwork to help individual members become supportive of one another. You can't do it all
by yourself, you know, so it is essential to get them to buy
into your program. Be the role model and the energizer to
set everyone in motion.

Build purpose and rewards to motivate your staff, and
be fair and consistent to win and keep their loyalty. Invite
suggestions and observations without displaying doubt

in your own judgment. Create transparency to make them feel included in your plans; but do not reveal so much as to make it difficult for you to make changes without engendering confusion.

Above all, be flexible enough to adopt necessary changes without appearing to second-guess your overall direction.

Synopsis
Charge your batteries and be ready to take the extra step.

Use Common Sense
A hammer has (at least) three surfaces.

Common sense can be summarized as recognizing what has worked in the past and still does. It consists of avoiding fairly obvious errors in favor of tried and proven measures to fairly ordinary problems. It also means that when one attempted solution (no matter how popular) doesn't work, consider switching to another alternative. Remember that most hammers have three functional sides: one for whacking nails, the second for yanking them out, and the third for grasping the handle.

What this means to women is that the qualities and characteristics commonly attributed to males and females can work as anticipated, in the opposite manner, or in completely unexpected ways. Watch what your colleagues and supervisors do differently: observe what works, the contexts in which they work best, and what doesn't. Adopt new perspectives for analyzing problems by transcending gender stereotypes; focus on what suits your style as an individual and as a woman.

Examples:
1. Refer to both males and females when citing good advice and positive examples.

2. Be as active as the men and women around you in taking the lead, offering suggestions, and (when appropriate) disagreeing with someone else's ideas.
3. Maintain an aura of courtesy and consideration; these qualities are by no means antithetical to confidence and strength.
4. You cannot lift the yoke of gender-based attitudes alone, but you can toss it off your shoulders by demonstrating your confidence, competence and professionalism.

Synopsis
Cross them up; think like a person, not a gender.

Summary
Pit your strong points against your competition's weaknesses. An important part of this approach is to let the principles of yin and yang work for you by keeping the competition off balance.

12. Coping
"If it's not to your advantage, don't advance." – Sun Tzu

In Sun Tzu's time, fire was a major weapon between warring factions. Today (terrorists and gangsters aside), attacks are usually more restrained. In the work environment they consist primarily of subtle threats and innuendo. Most often, the nastiest intrusions are launched behind your back.

A woman needs to be vigilant of the types of threats that can pester, or outright damage, her career. In fact, the only people who are largely immune to such attacks are those who remain so far beneath the radar they attract little attention. As with most threats, an understanding and awareness of the dynamics, and how to cope with them, lead to an ability to diffuse (or reverse) them. While not advocating unjust attacks against colleagues or competitors, in rare but volatile circumstances a preemptive move may be justified.

For a woman, the bottom line is that you don't need to be actively confrontational to draw attacks, merely perceived as a threat to someone. The attacker may target you personally, a proposal you have made, or an accomplishment. When this occurs, do not expect the Goddess of Fairness to intervene on your behalf: it's up to you to protect yourself. Which means standing up for you and your ideas, and emerging with a minimum of damage.

Attacks are either directed toward a person (group, etc) or an idea (suggestion, etc.). Let's take a look at the differences between these, and how to deal with them.

The Nature of Attack

"If you're not in danger, do not fight." – Sun Tzu

Paranoia doesn't necessarily mean that the threats that you perceive are imaginary. Within a competitive work environment, employees at all levels constantly jockey for position. When the competition is another company, political candidate, etc., you are dealing with an entirely different dynamic. We'll begin with the most immediate challenge a woman may be called upon to face: that of smiling colleagues plotting behind your back to push you aside.

Say you come up with a good idea: you research it, justify it, and begin to sell it to your boss. If he likes the concept, it will probably be revealed to certain of your colleagues for their input, perhaps at a department meeting.

Right away, you get a sense of who your supporters and detractors are likely to be. The arguments against your plan will identify how reasonable, illogical, or otherwise oriented they are. We'll get to how to deal with these in the following section; for now, let's focus upon identifying the nature of whatever objections you can expect to face.

Most likely, your opponents will attempt to build a case against your concept. If this tactic fails, they may shift gears to attack the proponent of the concept — you!

If either attack proves successful, your proposal will be doomed, and your career may be dealt a crushing blow.

Common objections against an idea:
1. It's been tried before and didn't work.
    [The circumstances were different, and it wasn't done the way I am proposing.]
2. It's never been done/we don't do things that way.
    [I'm sure we aren't all opposed to change for the better.]
3. Who knows what might happen?
    [Given the control factors I included, the probable results are more predictable than doing it the other/old way.]

Common objections against an individual:
    [push their objections back toward your proposal.]
1. She lacks experience.
    [What part of my proposal do you feel shows a lack of experience.]
2. She doesn't understand/realize how things are done here.
    [Having noted that things are done efficiently around here, my proposal appears to fit this pattern.]
3. We don't know if she can pull it off.
    [My proposal is straightforward and leaves little room for error.]

Synopsis
Line up support and isolate your enemies.

Negating an Attack

The best way to negate a threat is to diffuse it before it builds momentum. Having identified your main opponent, you might try co-opting him.

Example:

"Hi, Todd, I was thinking about your objections to my plan at the meeting and was wondering if we could discuss them together. I'd like to gain the benefit of your thoughts and experience."

Todd is likely to respond by:

1. agreeing to discuss, at which point you have taken a step toward winning his cooperation. Unless he is adamantly against your idea (or you), a compromise may be possible. If so, you may have co-opted a strong opponent to your side.

2. refusing (always possible), in which case you could approach your boss by saying something like, "I recognize that Todd is opposed to my idea. But when I tried to ask him more about his views, he seemed reluctant to discuss them with me. What do you recommend?"

3. politely putting you off by claiming to be busy, and so on. Which points you back to example 2) above.

OK, let's say that Todd is too pig-headed to be co-opted by your subtle touch. Now it's time to fight fire with a little heat of your own. But instead of a male-type head-to-head confrontation, you employ the feminine touch.

Example:

You tell your boss, "I've been trying to win Todd over to my proposal, but even after addressing his specific objections he still won't budge. Do you think it might be something personal?"

Naturally, your boss will reassure you that this is not the case, but it does serve to plant the possibility in his head. And it could deflect any negative comments Todd may make about you. At the same time, you've left the door open to Todd if he should decide to lighten up or cooperate with you.

Personal attacks can get ugly, and you must deal with them effectively.

First, you need to know that you are being attacked. Second, by whom: Todd, for example, or someone else? For this you need a confidant, someone who will let you know what's going on behind your back. Next, it wouldn't hurt to have an ally or two for support.

If you think something like this couldn't possibly happen to you, think again! Because the more successful you become, the higher you rise up the corporate ladder, the greater the possibility that you will be targeted by a jealous colleague/competitor.

To carry this to an unlikely but possible extreme, remember that if someone lies about you in a manner that could damage your career, they may be legally liable. This is not to suggest that you go running to the first lawyer you can find. Just mentioning the possibility, or having someone relay the message, is likely to pull them up short and stop the blabbing.

In addition, companies are particularly sensitive these days to any improper comments that demean female employees. Don't worry about hurting someone who has been trying to give you a bad time, because s/he does not deserve the sympathy. As for making an enemy, recognize that you already have one and are now dealing with them. When another person forces a win/lose situation on you, you don't have to be the loser.

Synopsis
Hoe your garden before the weeds take over.

Attack when Advantageous
Once you gain the advantage, do not waste it.

When faced with the choice of fending off a threat or going on the offense, Sun Tzu advises us to choose the latter only when our chances of succeeding are high. If the circumstances do not seem optimum, wait until conditions (those you can influence and the ones beyond your control) favor your side. Avoid reacting out of anger, resentment, or for that matter, unwarranted optimism. Since women are by nature more cautious than their male counterparts, let this dynamic work for you by reducing the probability of failure and increasing your odds of success.

Sun Tzu's references to troops translate readily to important resources. Just as you wouldn't pour your last cup of drinking water on a burning house, valuable resources (staff, budget and time) should not be wasted. Say one task requires two people for two weeks, and another will occupy the same two workers for six months: you would assess the priorities of the tasks before assigning your human resources. If both were equally important, you'd probably want to address the two-week job first. But if your two workers would only be available for half the time to get the job done, you'd think twice before beginning it.

Assessing the needed resources and the schedule against priorities and value (advantages of completing the job) enables you to make intelligent decisions as to when — or if — to launch a given task.

Example:
You've been thinking about making a push for a managerial position. When you hear that one of the managers is leaving (or being moved up the ladder), you arrange to meet with her. You prepare a few questions and solicit her advice as to your qualifications. Your strategy is to gain her support, perhaps even a recommendation, and to line her up as a potential mentor.

Synopsis
Recognize the moment and rise to the occasion.

Defend when Attacked

Create the appearance of an unassailable position.

Naturally, it makes good sense to set up a solid defense before you fall under attack. But for those of us who work in civilian life, too much defensiveness can isolate us and waste our creativity, energy and resources.

Governments struggle with the question of where to draw the line between vital defensive weapons and simply tossing money down the drain. Here again we need to find a balance (yin and yang) between appearing strong and building barricades against Don Quixote's windmills.

Sun Tzu advises us to be ready to cope with adversity through activity and deception. Appearing formidable and resourceful will discourage many forms of potential threat. Being ready means watching the signs without focusing on unnecessary fortifications. This will enable you to defend aggressively if an attack does come, giving you the time to mobilize additional resources as needed.

Nations devote billions to anti-missile and other defensive hardware. How much of this is practical, cost-effective, or actually needed? Clearly, political considerations drive much of this activity, with one party trying to show the public they are more concerned with the nation's safety than their opponents. Personal interests also enter the playing field as well-connected corporations are awarded highly profitable contracts. When the feared attacks fail to materialize, spending advocates claim that their policies and actions served as a deterrent, while their opponents cry "Waste!" History and retrospect have shown both positions to possess an element of truth. The key question is, how much safety and how much waste?

You, as an individual and a woman, cannot afford to squander your precious personal and professional resoures. There is only so much time in a day, so much energy to burn, and so much credibility to spend in completing your assignments and pursuing goals. At the end of this and every other day you need to reassess how much of what is available to you is being efficiently and effectively spent on fire prevention or wasted on exaggerated threats:

1. how real (probability of occurrence) and potentially damaging is the threat, and
2. how difficult/what resources would be needed to cope with it if it actually materialized.

As is so often the case, a few ounces of strategic prevention may be equal to a ton of wasted effort.

Your strategy for defending against a personal attack depends on:

1. whether it is subjective or backed by evidence;
2. who is attacking you, and
3. how potentially harmful is the threat?

Evidence based:
If someone is attacking a mistake you made, own up to it in a matter-of-fact manner.

Examples:
1. "As Fran has pointed out, the program still has a few bugs. For the record, we've been working on them for several days and expect to have them resolved within a week of the planned release."
2. "Our group budget, for which I accept the majority of responsibility, has been stretched beyond expectations. The reasons, not all of which could have been predicted by us, are in the process of being identified and documented."
3. "The problem, which we fully acknowledge, is actually less severe than feared by several observers. Let me elaborate, ..."

Subjective

Innuendo and accusations don't sink battleships, but they have damaged if not destroyed their share of individuals, and vindication often comes too late to save careers and reputations. Even unsupported allegations can poison a reputation and leave a taint long after the allegations are proven false. Ask any high school girl whether nasty whispers still have their age-old power. Or ask women who compete in corporate life. A few innuendos about alleged liaisons around the office, echoed by knowing nods, can create an ugly image.

Why do these outrageous situations continue to arise? Jealousy, or misdirected anger? More importantly, how can you defend yourself if they occur? Consider Lana's story, true except for the names:

Janna has been a district sales manager with GML Corp for six years, Lana for only three. Yet Lana was recently promoted to regional director, leaving Janna in the dust. Since they were the only female sales managers, their careers have been closely watched and compared.

Janna and her male colleagues are well aware of Lana's exceptional performance. Nevertheless, Janna and perhaps one or two of the guys feel overlooked, threatened, and generally unhappy that a woman with limited seniority has leapt ahead of them. Thus it began:

Janna: "You know, as a woman I hate to say this, but I wonder who Lana may have, you know, played up to in order to get chosen for regional."

Jim: "Well, I'm not going to mention any names, but I hear she may have been messing around with a certain exec."

Janna: "I'm sorry to hear that, but I have to admit it really doesn't surprise me."

Jim: "You know what they say — where there's smoke, there's fire."

Janna: (to another manager) "Hey, Will, did you hear what Jim said about Lana? Not that I would accuse her myself, mind you, but it does kind of make you wonder, doesn't it?"

Like most rumors, this one didn't need much fuel to spread like fire.

Eventually, Lana found out about the rumors through a fair-minded colleague and immediately set about dousing the flame. She began with an appointment with her boss.

She explained the situation openly and honestly: "Lew, I'm afraid there may be a potentially harmful rumor going around which could affect us both and I'd like to discuss it with you."

She suggested a strategy for dousing the rumors: "I'd like to put together a little presentation detailing sales performance district-by-district, giving credit to the better units and putting a little pressure on the underperformers. The tone wouldn't be critical, per se. The presenation could help identify problems and invite ideas about how to fix whatever may be wrong."

Lana's boss appreciated this disarming and productive approach and agreed to her suggestion. With her way clear, Lana drew up a few basic and non-critical presentations. She assigned Janna and Jim major roles in the preparation. To her replacement as district manager she provided clear and detailed guidance about the ins and outs of her former clients, expressed confidence in his ability to maintain and grow the district, and offered ongoing support: "Will, I know it can be hard to maintain an over performing district but we chose you because we felt that you could do a great job. So be sure to let me know whenever you want to discuss anything that concerns you."

Let's sum up Lana's response to the attack:
1. She alerted her boss to a potential problem and obtained his support;

2. She used simple facts to illustrate to Lana and Jim why she had been promoted. By involving them in the preparation, she gave them the opportunity to see that which was indisputable without embarrassing them. The same simple facts vouched for her performance to others in the department and in the company without singling out any one stellar or lesser performance;

3. She demonstrated her awareness of the performances in the department, subtly suggesting that Janna and Jim might wish to adjust their focus and concentrate on their results;

By using the presentation as a vehicle for discussion on what works and how problems can be identified and fixed, she expressed willingness to lead and assist so that all the associates could succeed.

And she accomplished all this without asking for her boss's intervention or burning any bridges. No wonder Lana's star is rising!

Synopsis
It isn't paranoid to prepare for an attack, or to defend yourself when it occurs.

Survive and Thrive
The higher up the corporate pyramid, the fewer the desks and tougher the competition. Within the corps of professional workers and middle management, the rivalry is generally less intense. Nevertheless, virtually every professional environment — from teaching to corporate life — presents its challenges.

The message to carry away with you is that it absolutely, positively isn't enough to do a good job; other issues can and will enter into the fray, making vigilance a must. Is it a bummer that office politics and personal agendas may complicate a woman's working life to a greater degree than that of men? Sure it is! That's why we'd like to help you deal with different kinds of potential annoyances, to make them a smaller part of your concerns.

Coping serves as a prerequisite to survival. It means dealing with problems of all magnitudes to keep them from derailing your train. With regard to threats, this is the minimum you need to achieve; if you fail to cope, survival becomes a dead issue.

Surviving means that you're coping well enough to keep the train running smoothly. With regard to your career, this is the minimum requirement to enable (but not guarantee) success. Nor does survival automatically equate to happiness, fulfillment and satisfaction. Frustrations and stress are always ready to build barriers across your chosen path.

Thriving is, of course, the goal upon which your career objectives are based: the train is running smoothly, on or a bit ahead of schedule, and you are enjoying at least a major portion of the ride. The reality of your career path, and the way you feel about it, are in synch. Yin and yang are in balance. Good for you!

Synopsis
Anticipate the worst, accept the best, and do what's right for you.

Summary
Unpleasant though it may be, time aside for dealing with the most disagreeable situations is justified by the damage to which you, and any other woman, may be vulnerable.

Why would someone want to hurt you? The reasons range from jealousy to fear of competition, callousness, or plain and simple nastiness. Most people aren't like that, but remember, it only takes a single driver running a red light to cause a terrible accident. Stay vigilant.

13. Networking

"Advanced information cannot be derived from the super-natural, deduced from random events, or conjectured from the sky; it must be obtained from living people." – Sun Tzu

Given the explosion of communications devices and methodologies that define our new millennium, it is difficult for us to imagine how limited information flow must have been in Sun Tzu's time. Back then, the only available resources for communicating with people in other areas and obtaining information was to visit them (a journey that could take weeks, or even months) or to send someone else. For many types of information, it was crucial to speak with someone familiar with your area of interest. Spies were essential to generals and rulers who wanted to invade another realm, and to avoid being invaded.

In modern warfare, as well as political and business practices, the spying profession continues to flourish. Most of this is performed sub rosa, but our newspapers and electronic media expose such activities nearly every day. Officially, corporations deny the practice, but many (if not most) employ "informants" within and outside their organizations.

Sun Tzu defined five kinds of spies:
1. Local: someone from your own territory.
2. Internal: a mole inside the opposition's organization.
3. Counterspy: a double agent working for the opposition.
4. Disposable: a spy whose loyalties are suspect and cannot be trusted, who is provided with false information to leak to the opposition.
5. Valued: a skilled undercover spy who can penetrate the opposition's information network.

He further cautioned that only a wise and subtle leader is able to use spies effectively. Otherwise she may be misled by vague, incomplete, or false information.

If you're wondering why we have been talking about spying under the title of networking, it is because we consider networking to be the practical and moral equivalent in today's business and social structure. For most of us as individuals, employing spies would be difficult, unrealistic, and ethically untenable. Fortunately, networking offers many of the same advantages within a pragmatic and accepted framework.

Why is this important to women? Because despite the good old boys' network long dominated by males, the majority of women are better listeners and thus superior communicators. These qualities make networking a potentially effective tool in a woman's arsenal.

Understand the Process

Networking is a systematic approach to a random process.

Networking is the art of sharing information. It is accomplished by gaining and maintaining access to people who can be helpful to you. Successful networking can facilitate your job search, assist your business, and augment your career in a number of ways.

Something else to understand about networking is that there are diverse ways to practice it. Various experts define the process differently, according to their preferences and what has worked best for them. Most agree about the things you need to do, and those you should avoid doing, to be successful. We will explore the most common winning formulas and how you can make them work to your advantage.

The history of networking is that of friends and acquaintances getting together to discuss topics of mutual interest. These conversations gradually expanded to include the

friends of friends and colleagues' colleagues. As more potential networkers jumped on board, the process began to
formalize into a career tool with rules and etiquette.

Instead of giving you a long list of do's and don'ts, we'll
boil them down to a few basic principles.

1. Be upfront: When you make a contact, let them know
   your purpose (information, job seeking, whatever).
   Never misrepresent your agenda.
2. Be concise: Prepare notes on the topics you want to cover.
3. Be brief: Keep it down to ten or fifteen minutes, unless
   a longer period is truly needed. Be considerate of the
   other person's time and interest.
4. Be helpful: Make an effort to provide some information
   that might be of interest to your contact. Be sure to follow up with any promised or requested information.

Synopsis
Follow networking etiquette.

Create Alliances

*"Enlightened rulers and perceptive generals who
are able to get intelligent spies will invariably
attain great achievements." – Sun Tzu*

Unless you are an insurance salesperson, employment
agent, financial advisor, or for other reasons feel the need to
develop an extensive list of contacts, target your potential
networking partners sparingly. Effective networking can
take a lot of time, and you have none to waste. Nor do the
people with whom you network.

Specifically, you want to be in touch with those who may
have something of value to offer you, and who may be interested in helping you. Otherwise why bother?

Understand as well that the networking process has
been overused by self-indulgent and inconsiderate indi-

viduals out on fishing trips (without a specific agenda) who abuse the system. That's why you need to be well organized and succinct when you contact someone, especially for the first time.

The bottom line for assessing a potential ally is whether they believe that you may also be of use to them. After all, networking (like most forms of communication) is a two-way street. So try to represent yourself as someone who the other person will value.

Now, if you are in the insurance, employment, finance, or another business that relies upon a detailed database of existing and potential clients, you will need to build an on-line or other form (perhaps index cards) of contact information that can conveniently be accessed and modified. You would probably want to list key dates (expirations, birthdays, and other reasons to contact) into a tickler file, one that automatically reminds you of a forthcoming event a day or two earlier.

Remember that alliances are based on mutual interest. The most successful of these are usually made up of companies and/or individuals who serve the same client base in a non-competitive manner.

For example, two real estate agents who focus on different markets (such as commercial and residential) may have common ground for cooperation. Likewise, financial analysts who specialize in mutual funds, derivatives or bonds might team up to offer a full package to potential investors.

As an individual employee, an overlooked type of alliance might be forged between two or more people with complimentary skill sets. If, for example, one is better with presentations and the other with operating systems, they might combine to create an appealing and expert technical presentation. Even offering a more experienced colleague your help with some of their mundane tasks in exchange for their insights and contacts could create a win-win alliance.

Look for colleagues with recognized skills, talents and influence, and ask yourself what you might have to offer them. When you approach them, tell them that you admire their ability, especially if it's one you would like attain or improve, and ask their advice. Breaking the ice in this manner may give you the opportunity to get to know them and identify what they might admire in you.

Cautionary note: both men and women may be suspicious of your intentions, so express yourself clearly and without a hint of innuendo. Misunderstandings at this stage can become embarrassing and damaging.

Synopsis
There are people out there who may need you as much as you need them. Meet them, cultivate them, and share with them.

Build Your Network
Lucky people are those who are ready.

The two main reasons for networking are to keep up with significant events where you work, and to explore the scene on the outside. So build your network with reliable sources in both arenas. Even if you aren't actively looking for a new position, you should always be prepared for opportunities when they arrive. The best way to do this is to proactively share information and, of course, keep your ears wide open.

Networking can help you in a number of ways. For example, learning about a new opening or project may enable you to get your foot inside the door before most of your competitors have heard about it. Conversely, one of your contacts might refer a great candidate to you for a position you need to fill. The value of information is well known, and networking can be an excellent resource.

What does it take to network effectively? It helps to have good communications skills, a sense of purpose, a pleasant demeanor, and a solid ego to withstand rejections. You'll also need to invest the time and effort to get started and keep the ball rolling.

A few tips:
1. Make a list of people to contact during the current week.
2. Set aside a period of 15-20 minutes for phone calls every day.
3. Fill in spare minutes (like a coffee break) to make a call or two.
4. Try to schedule lunch with a valued contact at least once a week.

Before calling, jot down a topic or tidbit of information like ly to interest the person you are contacting. Bear in mind that men typically need a rationale to justify a phone call ("The reason I called was to ..."), whereas women often call just to say hello.

Let's face it, successful networking requires hard work and persistence. It also takes a bite out of your clock, so plan and prepare your strategy accordingly. The art of net-working is a give and take process in which your contacts are willing to provide relevant information and introduce you to additional contacts. They, on the other hand, are counting on you to do the same for them.

When you call or email a new referral, identify the person who referred you and your reason for contacting them.

Example 1: "Charlie Morgenstern suggested I contact you egarding your extensive background in telemetric wire-less remote monitoring. This is a field in which I am very interested..."

If you are contacting someone without a referral, be sure to open with an interesting sentence that may capture their attention.

Example 2: "I read your interview in Business Week and
   sincerely hope you will not mind this inquiry. My reason for contacting you is, quite frankly, that I am very
   interested in working in a company like yours. If you
   decide to spare me a few minutes, you may find that my
   skills and experience fit in quite well..."

Without a personal referral, the probability of eliciting a
favorable response is probably small, but reaching out to a
key individual at the right time can sometimes beat the odds.

If your contact agrees to meet you in their office or a
neutral setting, perhaps for lunch, keep the words "calm"
and "courteous" in front of your mind's eye. Watch how
they react to you — especially their body language — to
be aware of their interest in what you are saying. Let them
determine the level of formality or informality with which
they seem most comfortable, but do not allow the conversation to progress too rapidly into personal matters. Basically,
use your common sense, listening and observing skills, and
intuition to guide you.

Networking guidelines
1. Tact: There's a time and place for offering your opinion
   or your resume.
2. Patience: Contacts may take months or years to bear
   fruit, if ever.
3. Purpose: Have a specific objective for each contact.
4. Clarity: Express yourself clearly and succinctly.
5. Creativity: Seek ways to tie their interests to yours.
6. Attitude: Maintain a positive approach.
7. Sensitivity: Listen carefully for any signs of impatience,
   irritability, or other negative flags.
8. Relevance: Offer advice and information that may be
   useful to the other person.
9. Record: Create a record of your contacts and conversations.

10. Respond: Always return calls and keep your promises.
11. Courtesy: Remember to say please and thank you.
12. Fairness: Never ask for inappropriate information or favors.
13. Honesty: Avoid untruths, exaggerations and innuendo.
14. Discretion: Keep negative and personal issues to yourself, and stay away from topics like politics, ethnicity and religion.

Synopsis
Don't let anyone waste your time, and don't waste theirs!

Share and Beware
*"Search for enemy agents who have
come to spy on you." – Sun Tzu*

For practical purposes there are two essential types of knowledge: knowing the subject, and knowing where to locate information about it. Depending on the nature of the subject matter, your research may involve looking things up on the web or certain documents, or knowing the right people to ask. The broader your network, the better your chance of finding the right resource.

Nevertheless, you are advised to choose your contacts judiciously in order to protect yourself and your time.

Your time
The time issue will become obvious when people begin to contact you in ever increasing numbers. They may appeal to your good nature for time-consuming favors that offer little possibility of reciprocity. If this sounds cold or callous, think again. The bottom line is that you cannot possibly help everyone, and so you'll need to decide on your criteria and priorities.

Many people are flattered by being asked for their opinions or personal assistance. For better or for worse, women are often seen as softer touches than men, and are thus more susceptible to such requests. So classify and prioritize the calls and emails you receive into potential network partners (who may have something to offer you) and one-way contacts (who are unlikely to be of help to you). Then decide how much time you can offer them without negatively impacting your schedule and needs. Of course, there's nothing wrong with helping people, as long as you don't suffer as a result.

What kind of people do you want as network partners? Begin with those with some experience in your areas of interest. Also, men and women who are employed in companies or specialties you may wish to target may be sources of valuable inside information. If they view you as a potential resource, so much the better.

Yourself

Aside from wasting time, the two most common mistakes that people make in networking are using large chunks of the workday to pursue their personal interests, and getting involved with the "wrong" people.

1. Integrity: Whether you find networking fascinating, engrossing, tedious, or somewhere in between, remember that the time taken from your working schedule is paid by your employer. We're not talking about a few minutes here and there, but rather blocks of hours. It isn't ethical to abuse this trust, and doing so can land you in a world of hurt. In addition, be aware that many companies monitor their employees' email!

2. Questionable contacts: Not everyone is forthright or well intended. In fact, some are dishonest about their reasons for wanting to network with you. Consider the

following examples, of which only the names and a few minor details have been altered.

Example 1: Roxanne joined the ABC insurance company as a claims adjuster. Betsy, the only other female adjuster, befriended her almost immediately, offering to introduce her around and help her get started. Roxanne was delighted to have Betsy's help and invited her to lunch to show her gratitude. Over time they continued to go out to lunch once or twice a week, usually at Betsy's suggestion.

When Roxanne's first quarterly review was held, her supervisor found fault with her performance, much of which had been influenced by Betsy's counsel. Roxanne came to realize that Betsy's objective had been to ensure that the only other female adjuster in the office didn't outshine her. To that end, she had fed Roxanne some misleading advice.

Roxanne learned her lesson without suffering too much damage. She started bringing her lunch to work with her, found a few other colleagues to network with, and earned much more favorable performance ratings thereafter.

Example 2: Linda was in the upper echelon of her company's sales representatives. Several of her colleagues, and a few outside contacts, occasionally asked her for advice. She did her best to help them without seriously impacting her time.

One day Mike, a sales rep from another company, made an unusual request: he asked Linda if he could accompany her on a few sales calls in order to "learn from the master."

Since Mike represented a product line that was not competitive with Linda's, she saw no reason to refuse. After all, Mike was pleasant company and his apparent admiration of Linda's professionalism was rather flattering.

Linda she thought no more of it until several months later, when she learned that Mike had moved to a another company and was now a direct competitor. The shocker was that he had called on her clients, spread several fabrications about Linda's company, and tried to steal her accounts. Whether Mike succeeded in luring any of Linda's clients is rather less important than whether Linda allowed his dishonesty to cripple her performance. She may have taken a hit or two, but she learned, moved on, and is still as successful and respected as ever.

The toughest side to networking is getting the ball rolling. Once you've gained some momentum, keep it going, because it's always hard to start all over again. Part of this is letting go of errors of judgment in your networking. Don't waste time bemoaning those instances in which you were beguiled. Learn from it and move on.

Synopsis
Maintain momentum by cultivating promising contacts and pruning your list.

Read the Handwriting on the Wall
"The means by which progressive rulers and perceptive generals moved and conquered was advanced knowledge." – Sun Tzu

Nothing happens in a vacuum. This applies to forthcoming and existing projects, staff and budget decisions, and the potential opportunities and pitfalls that surround you. The way to anticipate events that can affect your employment status is to keep your eyes and ears open. Narrowly focusing on your job while ignoring the handwriting on the wall is like walking through elevator doors without checking to see if the elevator is actually there.

When you extend your antenna beyond the limits of your cubicle or office, you'll be amazed at how much knowledge and speculation is circulating around you. Naturally you'll need to separate the facts from fiction:

Speculation is easily exposed by subjecting it to reason: does it make sense; is there evidence to support it; can it be confirmed; who is the source?

Facts are verified in a similar manner: they appear to fit a pattern of what you already know and seem reasonable. However, they may be difficult to confirm. Rule of thumb: if two or three usually reliable sources provide the same information, take the possibility seriously.

What sort of facts should you pay attention to? Details that could affect your job or status, such as mergers, significant staff and management changes, and budget cuts can be important to you. Such changes may open potential opportunities, or the need to update and circulate your resume.

Your networking contacts can help you to a point — it's up to you to use the information you receive to your advantage.

For example, say your employer is merging with another large company. Which of the two merging entities is likely to emerge on top (yours or theirs)? Does the other group have a department similar to the one you work in? How does your boss think your department will be affected? Bottom line: what are your prospects to survive and thrive?

Synopsis
Pay attention to what is in front of your nose.

Summary
Sun Tzu knew the value of information and advocated the use of spies to gather what he needed. Now-a-days millions of sophisticated professionals rely on networking, an organized supply-chain of key names, places and events of companies, industries, organizations, interest groups and activities.

Are you tuned in to a network that can help your career? If not, what are you waiting for?

Personal Profile — Catherine Huang
Being honest and trustworthy doesn't mean you have to let everyone know what you are thinking.

Catherine Huang runs an international import business she built from scratch over the past nine years. A former fashion designer, Ms Huang imports frozen fruit and vegetables from China, Egypt and Eastern Europe, and prepared desserts and breakfast foods from Belgium and the Netherlands. She was born in Taiwan to Chinese parents who fled Mainland China when Mao Zedung's Communist Party took control of the country.

I grew up in Taipei, where my father served as a military officer. My parents, who valued education, sent me to a very demanding school that taught classical Mandarin Chinese* and an unadulterated view of China's vast history.

Mom and Dad brought my two younger brothers and me to California when I was in my teens. Initially, the entire family had to depend upon my limited English, although they — especially my brothers — picked it up quickly. Like so many immigrant Chinese families, we opened a restaurant where I worked after school and did my homework in between waiting on tables and washing dishes.

Eventually I graduated from the Los Angeles Fashion Institute, and my brothers earned their degrees at UCLA. I began my career in the fashion industry in LA, moving on to Chicago and eventually New York City (where I was to meet my future husband on a blind date).

---

* The Mainland Chinese government altered the language taught in schools into a "simplified" version, and chose to emphasize or deemphasize various historic accounts to support their political agenda.

After years of gainful and, occasionally, creative employment, I set my past profession aside and jumped feet first into the world of business. Food importing is an overwhelmingly male-dominated industry, one in which I had absolutely no prior experience or personal contacts. Why did I choose it? Pretty much by chance and circumstance, and I was ready for a new challenge. Sure, it helped to have a supportive husband, but he had limited time to spend on my new enterprise, and I had to find my own way.

Now, some nine years later, I'm still struggling to expand my business, but I figure that's the way it is supposed to be. After all, a free lunch never tastes half as good as the one you have to earn.

Growing up within a Chinese culture, I have been aware of Sun Tzu's concepts most of my life. As a result, his advice seems more intuitive and common sense than revelation to me. It's hard to say exactly to what extent he influenced my approach to life and business, but I can point to a few working principles derived directly from *The Art of War*.

One particular example is Sun Tzu's insistence on personal integrity and the need for deception, which may, at first glance, seem to pose a contradiction of philosophies. To me they are completely separate issues.

"Integrity," pure and simple, means being honest and trustworthy. It's what I was taught at home and in elementary school, and my parents never let me forget it. Not a bad reputation to have in the world of commerce or elsewhere, when the people who have been dealing with you for years know that your word is good!

"Deception," on the other hand, is knowing how much information to give out and what to keep to yourself. It does not allow for lying, cheating, or other forms of dishonesty. Where I tend to use it most is when someone tries (successfully or not) to cheat me.

My tactic is to conceal how much I know about what they are up to; I just deal with it as effectively as I can, and let them go on thinking how much smarter than me they are. Of course, I never have anything to do with them again and will take whatever action necessary to protect myself. I may also share the details with a few close associates. Chances are they'll never know what I found out or did, and that's just fine with me.

Another of Sun Tzu's points is to thoroughly prepare well in advance of having to make an important decision. Before a business trip or meeting, I put together my game plan, build in an alternative or two, and am ready when an opportunity presents itself. This way I am rarely caught by surprise.

For those of you who have discovered Sun Tzu more recently, even if you only pick up a few pointers that help you in your career and life, I think you'll find it worth your effort.

— *Catherine Huang*

Index